Disaster Management and Sustainable Development

Disaster Management
and
Sustainable Development

By

Niranjan Sahoo
Associate Professor,
Department of Rural Management,
Xavier Institute of Social Service (XISS), Ranchi

New Century Publications
New Delhi, India

New Century Publications
4852 / 24, 1st Floor,
Ansari Road, Daryaganj,
New Delhi – 110 002 (India)

Tel.: 011-4101 7798; 4358 7398; 98112 66355
E-mail: info@newcenturypublications.com

www.newcenturypublications.com

Editorial office:
4800 / 24, Bharat Ram Road,
Ansari Road, Daryaganj,
New Delhi – 110 002

Reprint: 2025

ISBN: **978-81-7708-477-1**

First Published: 2019

Published by New Century Publications and printed at Milan Enterprises, New Delhi.

Designs: Patch Creative Unit, New Delhi.

PRINTED IN INDIA

About the Book

A disaster is an extreme disruption of the functioning of a society that causes widespread human, material, and environmental losses that exceed the ability of the affected society to cope with its own resources.

India's geo-climatic conditions—as well as its high degree of socio-economic vulnerability—make it one of the most disaster-prone countries in the world. Disasters are sometimes classified according to whether they are *natural disasters*, or *human-made disasters*. For example, disasters caused by floods, droughts, tidal waves and earth tremors are generally considered *natural disasters*. Disasters caused by chemical or industrial accidents, environmental pollution, transport accidents and political unrest are classified as *human-made* or *human-induced disasters* since they are the direct result of human actions.

The super cyclone in Orissa in October 1999 and the Bhuj earthquake in Gujarat in January 2001 underscored the need to adopt a multi-dimensional endeavour involving diverse scientific, engineering, financial and social processes. Need was also felt to adopt a multi-disciplinary and multi-sectoral approach and incorporation of risk reduction in the developmental plans and strategies.

Disaster Management Act, 2005 changed the definition of disaster which now includes not only the events emanating from natural and man-made causes, but even those which are caused by accident or negligence.

Prime Minister Shri Narendra Modi released the first Disaster Management Plan of India on June 1, 2016 that seeks to provide a framework and directions for government agencies for prevention, mitigation and management of disasters.

This book provides a detailed account of disaster mitigation and management policies and programmes in India, including the social responsibility of corporates.

About the Author

Dr. Niranjan Sahoo is currently Associate Professor in the Department of Rural Management, Xavier Institute of Social Service (XISS), Ranchi. He did his M.A. from Visva-Bharati Central University, Santiniketan, West Bengal; M.Phil. from BANISS (an ICSSR Institute), University of Indore; and Ph.D. from Magadh University. He also holds STC on Human Resource Management from IIT-Kharagpur.

He is the recipient of many awards, notably International Intellectual Achiever's Award conferred in Bangkok, Thailand (December 30, 2010); Visva-Bharati Snattakottar Award conferred by the then Prime Minister Shri I.K. Gujral (January 11, 1997); Rashtriya Gaurav Award (April 9, 2010) and Rajeev Gandhi Excellence Award (September 9, 2010). Recently, he was selected a member to the Apex Committee of Uranium Corporation of India Limited (UCIL), Department of Atomic Energy, Government of India. Dr. Sahoo has visited a few South Korean universities—viz. University of Suwon, University of Sinhan, Vision University, Hwashin Cybercity University and Chanbuk National University—as guest professor.

Dr. Sahoo has authored a book titled *Rural Development and Livelihoods in India* and co-authored another book titled *Demonetization, Digital India and Governance*. He has published a number of research papers in academic journals of repute including *International Journal of Multi-disciplinary Education and Research*, and *Journal of International Centre for Economics, Humanities and Management*, Bangkok. His areas of research interest include rural financial services, corporate social responsibility (CSR), entrepreneurship development, and human resources management.

Contents

About the Book..*v*
About the Author..*vi*
Preface..*xi-xiv*

1. Sustainable Development and Disaster Reduction **1-17**
1.1 Sustainable Development Defined
1.2 Sustainable Development Goals (SDGs)
1.3 Sustainable Development Goals Report, 2018
1.4 World Conference on Disaster Risk Reduction
1.5 Measures to Promote Sustainable Development
1.6 International Efforts and Co-operation

2. History and Case Studies of Disasters **18-32**
2.1 History of Disasters
2.2 Orissa Super Cyclone, October 29-30, 1999
2.3 Community-based Disaster Preparedness (CBDP)
2.4 Tenth Five Year Plan (2002-07) and Orissa Super Cyclone
2.5 Floods in Bangladesh

3. Pipeline Disasters and Oil Spills **33-39**
3.1 Pipeline Transportation
3.2 Oil Spills
3.3 Nigerian Dilemma

4. Climate Change and Disasters **40-63**
4.1 Climate Change and Human Settlements
4.2 Climate Change and Water Supply
4.3 Occurrence of Droughts
4.4 River Hydrology and Inland Flooding

4.5 Cyclonic Storms, Storm Surges and Coastal Flooding
4.6 Sea Level Rise
4.7 India's Approach to Climate Change
4.8 National Action Plan on Climate Change (NAPCC)
4.9 Eight National Missions of NAPCC
4.10 Other National and Sub-National Initiatives

5. Land Degradation and Droughts **64-87**
5.1 Land Pollution and Degradation
5.2 Consequences of Land Degradation
5.3 Strategies to Check Land Degradation
5.4 Droughts: Causes, Consequences and Remedial Strategies
5.5 Programmes for Treatment of Degraded Lands

6. Disaster Mitigation and Management in India **88-99**
6.1 India's Vulnerability to Natural Disasters
6.2 Disaster Management and Development Planning
6.3 Disaster Management Act, 2005
6.4 National Disaster Response Force (NDRF)
6.5 National Institute of Disaster Management (NIDM)
6.6 Twelfth Five Year Plan (2012-17) on Disaster Management
6.7 Disaster Management and Finance Commissions
6.8 Gujarat Institute of Disaster Management (GIDM)

7. Disaster Management Education 100-109
7.1 Role of Education for Disaster Preparedness
7.2 Disaster Management Education in India
7.3 National Disaster Management Training and Education for Disaster Preparedness
7.4 Public Education, Training, Rehearsal and Experience for Disaster Preparedness
7.5 Education, Experience and Risk Perception for Disaster Preparedness

8. Concept and Benefits of Corporate Social Responsibility (CSR) 110-129
8.1 What is CSR?
8.2 Main Features of CSR
8.3 Origins of CSR
8.4 Approaches to CSR
8.5 Principle of Triple Bottom Line (TBL)
8.6 Potential Business Benefits of CSR
8.7 Convergence and Transparency of CSR Practices
8.8 Globalization and CSR
8.9 Promoting CSR: The Challenges
8.10 Criticisms and Concerns

9. Corporate Social Responsibility (CSR) in India 130-143
9.1 Pressure for CSR
9.2 CSR and Disaster Preparedness
9.3 Relevance of CSR in India
9.4 Initiatives of Reserve Bank of India (RBI)
9.5 CSR in Companies Act, 2013
9.6 Some Concerns

10. Disaster Management in Select Countries 144-148
10.1 USA
10.2 Canada

10.3 Australia
10.4 South Africa
10.5 Japan
10.6 Brazil

Appendix 1: Classification of Disasters **149-151**

Appendix 2: Kerala Floods 2018: Damage and Need Assessments **152-168**

Bibliography **169-174**

Index **175-182**

Preface

Disasters are not new to mankind. They have been the constant, though inconvenient, companions of human beings since time immemorial. Disasters can be natural or human-made. Earthquakes, cyclones, hailstorms, cloud-bursts, landslides, soil erosion, snow avalanches, floods etc. are the examples of natural disasters while fire, epidemics, road, air, rail accidents and leakages of chemicals/nuclear installations etc. fall under the category of human-made disasters.

Disaster is an event or series of events, which gives rise to casualties and damage or loss of property, infrastructure, environment, essential services or means of livelihood on such a scale which is beyond the normal capacity of the affected community to cope up with. Disaster is also sometimes described as a catastrophic situation in which the normal pattern of life or eco-system has been disrupted and extra-ordinary emergency interventions are required to save and preserve lives and or the environment.

India's geo-climatic conditions as well as its high degree of socio-economic vulnerability, makes it one of the most disaster-prone countries in the world. A disaster is an extreme disruption of the functioning of a society that causes widespread human, material, or environmental losses that exceed the ability of the affected society to cope with its own resources. Disasters are sometimes classified according to whether they are *natural disasters*, or *human-made disasters*. For example, disasters caused by floods, droughts, tidal waves and earth tremors are generally considered *natural disasters*. Disasters caused by chemical or industrial accidents, environmental pollution, transport accidents and political unrest are classified as *human-made* or *human-induced disasters* since they are the direct result of human action.

A more modern and social understanding of disasters, however, views this distinction as artificial since most disasters result from the action or inaction of people and their social and economic structures. This happens by people living in ways that degrade their environment, developing and over populating urban centres, or creating and perpetuating social and economic systems. Communities and population settled in areas susceptible to the impact of a raging river or the violent tremors of the earth are placed in situations of high vulnerability because of their socio-economic conditions. This is compounded by every aspect of nature being subject to seasonal, annual and sudden fluctuations and also due to the unpredictability of the timing, frequency and magnitude of occurrence of the disasters.

The super cyclone in Orissa in October 1999 and the Bhuj earthquake in Gujarat in January 2001 underscored the need to adopt a multi-dimensional endeavour involving diverse scientific, engineering, financial and social processes; the need to adopt multi-disciplinary and multi-sectoral approach and incorporation of risk reduction in the developmental plans and strategies.

Over the years, the Government of India has brought about a paradigm shift in the approach to disaster management. The new approach proceeds from the conviction that development cannot be sustainable unless disaster mitigation is built into the development process. Another corner stone of the approach is that mitigation has to be multi-disciplinary, spanning across all sectors of development. The new policy also emanates from the belief that investments in mitigation are much more cost effective than expenditure on relief and rehabilitation.

Disaster management occupies an important place in India's policy framework as it is the poor and the under-privileged who are the worst affected on account of calamities and disasters.

On December 23, 2005, the Government of India enacted the Disaster Management Act, which envisaged the creation of National Disaster Management Authority (NDMA), headed by the Prime Minister, and State Disaster Management Authorities (SDMAs) headed by the respective Chief Ministers, to spearhead and implement a holistic and integrated approach to disaster management in India.

The Disaster Management Act, 2005, defines disaster as "a

catastrophe, mishap, calamity or grave occurrence in any area, arising from natural or man-made causes, or by accident or negligence which results in substantial loss of life or human suffering or damage to, and destruction of, property, or damage to, or degradation of, environment, and is of such a nature or magnitude as to be beyond the coping capacity of the community of the affected area".

The perception about disaster and its management has undergone a change following the enactment of the Disaster Management Act, 2005. The definition of disaster is now all encompassing, which includes not only the events emanating from natural and man-made causes, but even those events which are caused by accident or negligence. There was a long-felt need to capture information about all such events occurring across the sectors and efforts made to mitigate them in the country and to collate them at one place in a global perspective.

Prime Minister Shri Narendra Modi released the first Disaster Management Plan of India on June 1, 2016, that seeks to provide a framework and direction for government agencies for prevention, mitigation and management of disasters. This is the first national plan since the enactment of the Disaster Management Act, 2005.

Acknowledgements

I take this opportunity to place on record my deep appreciation and gratefulness to my mentors, colleagues and friends for their support and co-operation at every stage of the completion of this work. The whole process of project was guided and supported by many people. I sincerely thank all of them for their thinking, proactive attitudes and support. I am graciously owed for the success of writing this book which was an offshoot of sincere efforts channelled by Xavier Institute of Social Service (XISS), Ranchi. I greatly acknowledge my sincere appreciation to Fr. (Dr.) Alexius Ekka, SJ, Director of XISS, Fr. Pradeep Kerketta, SJ, Assistant Director of XISS, Fr. Xavier Soreng, SJ, Superior of XISS, and Fr. Marianus Kujur, SJ, Chairman of XISS

Governing Body.

As a colleague, I would like to acknowledge with gratitude for the luminary support all my colleagues provided me. I am deeply indebted to Prof. (Dr.) Himadri Sinha (HoD-RM), Prof. (Dr.) M.H. Ansari, Prof. (Dr.) K.K. Bhagat, Prof. (Dr.) Anant Kumar, Prof. (Dr.) Sant Kumar Prasad, Dr. Rajshree Verma and Dr. Sanjay Verma.

I am extremely thankful to Dr. Sarika R. Lohana, Mumbai without whom this book writing project would not have been completed. She is the only one who gave me an idea on how to write the book and this book is the outcome of her guidance.

My amorous and sentimental wife Prof. (Dr.) Sonia Sahoo and my pretty daughter Ms. Snehal Sahoo are the real inspirations for this effort. Though they are not currently staying with me but they have tacitly supported me without any complaints. I thank and love both of them for their meaningful contributions.

Finally, I would like to thank New Century Publications, New Delhi for bringing out this book on the day of my 18th marriage anniversary.

Ranchi **Professor Niranjan Sahoo**
October 1, 2018

1

Sustainable Development and Disaster Reduction

1.1 Sustainable Development Defined

The concept of sustainable development is inextricably linked with environment protection. It is a strategy for improving the quality of human life while living within the carrying capacity of supporting eco system. It is clear from the above definition that sustainable development entails commitment, protection and preservation.

Sustainable development is that which meets the needs of the present without compromising the ability of future generations. The heaviest burden in international economic adjustment has been carried by the world's poorest population. The consequence has been the considerable increase in human distress and over exploitation of land and water resources to ensure survival in the short term. Further, the desire to solve all social and human problems at one stroke has influenced the pace of development. Its reaction is visible in the atmosphere and in natural, physical and social environment. Awareness regarding the consequences of this event has developed into a consensus referred to as ecological concern.

Sustainable development advocates economic progress in an environmentally responsible manner. Sustainable development attempts to strike a balance between the requirements of economic development and the need for protection of the environment. It seeks to combine the elements of economic efficiency, inter-generational equity, social concerns and environmental protection. Although, the term sustainable development has many interpretations, it generally refers to non-declining human well-being over time.

Economic growth and development have to be guided by

the compulsion of sustainability, because no country can afford to ignore the economic as well as the environmental threat that a fast-deteriorating ecosystem poses to its economy. No country is immune to the reality of climate change, ecological degradation, depletion of the ozone layer and contamination of freshwater.

Sustainable development is a difficult balancing act in countries with low incomes. Society has to simultaneously accomplish three things with trade-offs: improve economic well-being with social justice for the present generation, yet manage with more restrained use of land, air, forest, energy, and water resources, and protect future generations. The choices are more difficult in developing countries because they affect people's livelihoods. Such a 'stewardship' to succeed therefore needs to respond to people's needs, share information on choices and costs, and ensure participation and ownership.

1.2 Sustainable Development Goals (SDGs)

On January 1, 2016, the 17 Sustainable Development Goals (SDGs) of the 2030 Agenda for Sustainable Development, adopted by world leaders in September 2015 at an historic UN Summit, officially came into force.

SDGs are the blueprint to achieve a better and more sustainable future for all. They address the global challenges, including those related to poverty, inequality, climate, environmental degradation, prosperity, and peace and justice. The Goals interconnect and in order to leave no one behind, it is important to achieve each Goal and target by 2030.

These new Goals that universally apply to all, countries will mobilize efforts to end all forms of poverty, fight inequalities and tackle climate change, while ensuring that no one is left behind.

SDGs build on the success of the Millennium Development Goals (MDGs) and aim to go further to end all forms of poverty. The new Goals are unique in that they call for action by all countries, poor, rich and middle-income to promote prosperity while protecting the planet. They recognize that ending poverty must go hand-in-hand with strategies that build economic growth and addresses a range of social needs including education, health, social protection, and job opportunities, while tackling climate change and environmental protection.

The seventeen goals are as under:

Goal 1: End poverty in all its forms.
Goal 2: Zero hunger.
Goal 3: Health.
Goal 4: Education.
Goal 5: Gender equality and women's empowerment.
Goal 6: Water and sanitation.
Goal 7: Energy.
Goal 8: Economic growth.
Goal 9: Infrastructure, industrialization.
Goal 10: Inequality.
Goal 11: Cities.
Goal 12: Sustainable consumption and production.
Goal 13: Climate change.
Goal 14: Oceans.
Goal 15: Biodiversity, forests, desertification.
Goal 16: Peace, justice and strong institutions.
Goal 17: Partnerships.

1.3 Sustainable Development Goals Report, 2018

According to this Report, a fast-changing climate, conflict, inequality, persistent pockets of poverty and hunger and rapid urbanization are challenging countries' efforts to achieve SDGs.

The Report found that conflict and climate change were major contributing factors leading to growing numbers of people facing hunger and forced displacement, as well as curtailing progress towards universal access to basic water and sanitation services.

For the first time in more than a decade, there are now approximately 38 million more hungry people in the world, rising from 777 million in 2015 to 815 million in 2016. According to the Report, conflict is now one of the main drivers of food insecurity in 18 countries. In 2017, the world experienced the costliest North Atlantic hurricane season on record, driving the global economic losses attributed to the disasters to over US$ 300 billion.

At the same time, the Report found that more people are leading better lives than they were just a decade ago. The proportion of the world's workers living with their families on less than US$ 1.90 per person a day declined significantly over the past two decades, falling from 26.9 percent in 2000 to 9.2 percent in 2017.

The under-five mortality rate dropped by almost 50 percent and in the least developed countries, the proportion of population with access to electricity has more than doubled between 2000 and 2016. However, in 2015, 2.3 billion people still lacked even a basic level of sanitation service and 892 million people continued to practice open defecation. In 2016, there were 216 million cases of malaria compared to 210 million cases in 2013 and close to 4 billion people were left without social protection in 2016.

SDG Report presents an overview of progress toward achieving the Goals, which were unanimously adopted by countries in 2015. As the global community moves forward to achieve the SDGs and address existing challenges, reliable, timely, accessible and disaggregated data is critically needed. This requires technology and innovation, increased resources and political commitment to build strong data and statistical systems in all countries.

1.4 World Conference on Disaster Risk Reduction

The World Conference on Disaster Risk Reduction is a series of United Nations conferences focusing on disaster and climate risk management in the context of sustainable development. The World Conference has been convened three times, with each edition to date having been hosted by Japan in Yokohama in

1994, in Kobe in 2005 and in Sendaiin 2015.

As requested by the UN General Assembly, the United Nations Office for Disaster Risk Reduction (UNISDR) served as the coordinating body for the Second and Third UN World Conference on Disaster Reduction in 2005 and 2015.

The conferences bring together government officials and other stakeholders, such as NGOs, civil society organizations, local government and private sector representatives from around the world to discuss how to strengthen the sustainability of development by managing disaster and climate risks. The Third UN World conference adopted the Sendai Framework for Disaster Risk Reduction 2015-2030. Previous conference outcomes include the Hyogo Framework for Action 2005-2015: Building the Resilience of Nations and Communities to Disasters in 2005 and the Yokohama Strategy and Plan of Action for a Safer World in 1994.

1.4.1 First World Conference on Natural Disasters in Yokohama, 1994: First World Conference on Natural Disasters in Yokohama, Japan, May 23 to 27, 1994, adopted the Yokohama Strategy for a Safer World: Guidelines for Natural Disaster Prevention, Preparedness and Mitigation and its Plan of Action, endorsed by the UN General Assembly in 1994. It was the main outcome of the mid-term review of the International Decade of Natural Disaster Reduction (IDNDR) and established 10 principles for its strategy, a plan of action and a follow-up. Furthermore, it provides guidelines for natural disaster prevention, preparedness and mitigation.

Ten principles of the Yokohama Strategy for a Safer World are as follows:

1. Risk assessment is a required step for the adoption of adequate and successful disaster reduction policies and measures.
2. Disaster prevention and preparedness are of primary importance in reducing the need for disaster relief.
3. Disaster prevention and preparedness should be considered integral aspects of development policy and planning at

national, regional, bilateral, multilateral and international levels.

4. The development and strengthening of capacities to prevent, reduce and mitigate disasters is a top priority area to be addressed during the Decade so as to provide a strong basis for follow-up activities to the Decade.
5. Early warnings of impending disasters and their effective dissemination using telecommunications, including broadcast services, are key factors to successful disaster prevention and preparedness.
6. Preventive measures are most effective when they involve participation at all levels, from the local community through the national government to the regional and international level.
7. Vulnerability can be reduced by the application of proper design and patterns of development focused on target groups, by appropriate education and training of the whole community.
8. The international community accepts the need to share the necessary technology to prevent, reduce and mitigate disaster; this should be made freely available and in a timely manner as an integral part of technical cooperation.
9. Environmental protection as a component of sustainable development consistent with poverty alleviation is imperative in the prevention and mitigation of natural disasters.
10. Each country bears the primary responsibility for protecting its people, infrastructure, and other national assets from the impact of natural disasters. The international community should demonstrate strong political determination required to mobilize adequate and make efficient use of existing resources, including financial, scientific and technological means, in the field of natural disaster reduction, bearing in mind the needs of the developing countries, particularly the least developed countries.

1.4.2 Second World Conference on Disaster Reduction in Kobe, 2005: The Second World Conference on Disaster Reduction conference was held in Kobe, Japan from January 18 to 22, 2005. This conference took on particular poignancy, as it came almost 10 years to the day after the Great Hanshin earthquake in Kobe and less than a month after the 2004 Indian Ocean tsunami. Japan's long history of severe natural disasters, prominence in international humanitarian aid and development and its scientific achievements in monitoring dangerous natural phenomena also made it a suitable conference venue.

The upcoming conference had not garnered much attention, but due to the December 26, Indian Ocean tsunami, the attendance grew dramatically and the international media focused on the event. Japan's Emperor Akihito opened the conference and welcomed 4,000 participants from around the world.

The World Conference adopted plans to put in place an International Early Warning Programme (IEWP), which had first been proposed at the Second International Conference on Early Warning in 2003 in Bonn, Germany.

The goal of the World Conference was to find ways to reduce the toll of disasters through preparation, and ultimately to reduce human casualties. Due to the proximity to the devastating Indian Ocean tsunami, developing a global tsunami warning system was high on the agenda. Other topics included:

- Pledges to reduce disaster damage.
- Healthcare after disaster.
- Early warning systems.
- Safe building standards.
- Agree upon cost-effective preventive counter measures.
- A global database on relief and reconstruction and a centre on water hazards.

The Pacific Rim Tsunami Warning system is an example of a cost-effective warning system; its yearly operating cost is approximately US$ 4 million. The yearly operating cost of a hypothetical global warning system is estimated at US$

30 million. This cost, compared to the international aid donations of nearly US$ 8 billion for the 2004 Indian Ocean earthquake and tsunami, clearly demonstrates the cost effectiveness of such a system.

A. Hyogo Framework for Action: The Hyogo Framework for Action (2005-2015): Building the Resilience of Nations and Communities to Disasters was an outcome of the 2005 conference. The Hyogo Framework (HFA) was the first plan to explain, describe and detail the work required from all different sectors and actors to reduce disaster losses. It was developed and agreed on with the many partners needed to reduce disaster risk—governments, international agencies, disaster experts and many others—bringing them into a common system of coordination. The HFA, which ran from 2005 to 2015, set five specific priorities for action:

1. Making disaster risk reduction a priority;
2. Improving risk information and early warning;
3. Building a culture of safety and resilience;
4. Reducing the risks in key sectors; and
5. Strengthening preparedness for response.

1.4.3 Third UN World Conference on Disaster Risk Reduction (WCDRR), 2015: The Third UN United Nations World Conference on Disaster Risk Reduction was held in Sendai, Japan from March 14 to 18, 2015, drawing 6,500 delegates to the conference itself and 50,000 people to the associated Public Forum. Sendai is the largest city of Miyagi Prefecture, in north-eastern Japan. It has a prominent status as it was hit by the Great East Japan earthquake, 130 kilometres from the epicentre. The conference included discussion of the aftermath of the Japanese response to the 2011 disaster and how Japan's early warning system can save lives when earthquakes and tsunamis strike.

The conference included an announcement of a US$ 4 billion fund to prepare for disasters over four years. Moreover, the conference coincided with Cyclone Pam hitting Vanuatu, and Vanuatu President Lonsdale urgently requested international assistance for his people. Thus, in 2015, many developed nations announced they would partner with smaller countries to prepare for future disaster relief operations.

The conference adopted the Sendai Framework for Disaster Risk Reduction 2015-2030. The Sendai Framework is the first major agreement of the post-2015 development agenda, with seven targets and four priorities for action. It was endorsed by the UN General Assembly in June 2015.

A. Sendai Framework for Disaster Risk Reduction 2015-2030: The Sendai Framework is a 15-year non-binding agreement which recognizes that the State has the primary role to reduce disaster risk but that responsibility should be shared with other stakeholders including local government and the private sector. It aims for the following outcome: "The substantial reduction of disaster risk and losses in lives, livelihoods and health and in the economic, physical, social, cultural and environmental assets of persons, businesses, communities and countries".

The Sendai Framework emerged from three years' of consultations and negotiations, supported and coordinated by UNISDR, during which UN member states, NGOs and other stakeholders made calls for an improved version of the existing Hyogo Framework, with a set of common standards, a comprehensive framework with achievable targets, and a legally-based instrument for disaster risk reduction. Based on the Sendai Framework, the first was held in New Delhi as part of Asian ministerial conference for disaster risk reduction. Member states also emphasised the need to tackle disaster risk reduction and climate change adaption when setting the Sustainable Development Goals, particularly in light of an insufficient focus on risk reduction and resilience in the original Millennium Development Goals.

The Sendai Framework sets four specific priorities for action:

1. Understanding disaster risk.
2. Strengthening disaster risk governance to manage disaster risk.

3. Investing in disaster risk reduction for resilience.
4. Enhancing disaster preparedness for effective response, and to "Build Back Better" in recovery, rehabilitation and reconstruction.

To support the assessment of global progress in achieving the outcome and goal of the Sendai Framework, seven global targets have been agreed:

1. Substantially reduce global disaster mortality by 2030, aiming to lower average per 100,000 global mortality between 2020-2030 compared to 2005-2015.
2. Substantially reduce the number of affected people globally by 2030, aiming to lower the average global figure per 100,000 between 2020-2030 compared to 2005-2015.
3. Reduce direct disaster economic loss in relation to global gross domestic product by 2030.
4. Substantially reduce disaster damage to critical infrastructure and disruption of basic services, among them health and educational facilities, including through developing their resilience by 2030.
5. Substantially increase the number of countries with national and local disaster risk reduction strategies by 2020.
6. Substantially enhance international cooperation to developing countries through adequate and sustainable support to complement their national actions for implementation of the framework by 2030.
7. Substantially increase the availability of and access to multi-hazard early warning systems and disaster risk information and assessments to the people by 2030.

1.5 Measures to Promote Sustainable Development

The role of economic instruments in financing and promoting sustainable development is well recognised. The following economic instruments can help achieve sustainable development through their influence on behavioural patterns leading to sustainable consumption and production in the economy:

1.5.1 Environmental Taxes: An environment or green tax is imposed on a product that damages the environment, in an attempt to reduce its production or consumption. Well-designed environmental taxes and other economic instruments can play an important role in ensuring that prices reflect environmental costs, in line with the *polluter pays principle*. Environmental taxes can be simple and efficient financial instruments for improving the productivity of natural resources. Environmental taxes on water and fossil fuels need not be part of the general revenues of the Government. Rather, they should be directly ploughed back into environmentally sustainable action on these fronts. Coal cess is a good example of environment tax imposed by Government of India in recent times, whose proceeds are channelled to the National Clean Energy Fund. Some benefits of environment taxation are enumerated below:

1. They provide incentives for measures that protect the environment, and deter actions that lead to environmental damage.
2. Economic instruments such as taxes can enable environmental goals to be achieved at the lowest cost, and in the most efficient way.
3. By internalizing environmental costs into prices, they help signal the structural economic changes needed to move to a more sustainable economy.
4. They can encourage innovation and development of new technology.
5. The revenue raised by environmental taxes can be used to reduce the level of other taxes. This could help reduce distortions, while raising the efficiency with which resources are used in the economy.

1.5.2 Subsidies: There has been a growing recognition of fiscal and environmental implications of the subsidy policies in energy, water and agriculture sectors. Most of these subsidies pose a threat to the environment. For example, in India, the subsidy on diesel has distorted the use of energy in transport and industrial sectors, and worsened the problem of hazardous air pollution. Similarly, excessive use of nitrogenous fertilizers and over-drawing of water, backed by subsidies at both Centre and State levels, is playing havoc with the sustainability of soil and water ecosystem.

1.5.3 Non-monetary Incentives: Non-monetary incentives are policy instruments that typically do not have a monetary value, but definitely have a financial impact that promotes sustainability. These incentives can be used as a bargaining tool by the Government to encourage conservation of resources in an economy. Activities such as those encouraging judicious use of water, planting trees, car pooling and avoiding use of plastic bags can be rewarded so that it encourages the practice, and acts as an example for others. Through the initiation of innovative policies and awards, the Government can provide recognition, which will encourage sustainable development amongst the citizens and the firms.

1.5.4 Eco-Industrial Hubs: An eco-industrial park (EIP) or estate is a community of manufacturing and service businesses located together on a common property. Member businesses seek enhanced environmental, economic and social performance through collaboration in managing environmental and resource issues. By working together, the community of businesses seeks a collective benefit that is greater than the sum of individual benefits each company would realise by only optimizing its individual performance.

The goal of an EIP is to improve the economic performance of the participating companies while minimizing their environmental impacts. Components of this approach include green design of the park infrastructure (new or retrofitted); cleaner production; prevention of pollution; energy efficiency and inter-company partnering. An EIP also seeks benefits for neighbouring communities to assure the net impact of its development is positive.

1.6 International Efforts and Co-operation

Environment has become an important issue in recent years. The Stockholm Conference on Human Environment held in 1972 and the Earth Summit held in Rio de Janeiro in June 1992 [United Nations Conference on Environment and Development (UNCED)] served as catalysts to focus the attention of mankind on the deteriorating quality of environment.

Global interfaces are gaining increasing importance in the field of environment. Environment is characterised by interconnectedness that transcends national/international boundaries and hence international cooperation and national efforts are seminally important to achieve the objectives of equitable access to clean air and water, adaptation and mitigation of climate change, conservation of biodiversity, sustainable forest management, safety in the management of chemicals, wastes and other hazardous substances.

1.6.1 Millennium Development Goals: In 2000, the 191 members of the United Nations, including India, committed to achieving 8 Millennium Development Goals by 2015. The 7th of these 8 Goals pertains to environmental sustainability:

1. **Eradicate Extreme Poverty and Hunger:**
   - Reduce by half the proportion of people living on less than a dollar a day.
   - Reduce by half the proportion of people who suffer from hunger.
2. **Achieve Universal Primary Education:**
   - Ensure that all boys and girls complete a full course of primary schooling.
3. **Promote Gender Equality and Empower Women:**
   - Eliminate gender disparity in primary and secondary education preferably by 2005, and at all levels by 2015.
4. **Reduce Child Mortality:**
   - Reduce by two thirds the mortality rate among children under five.
5. **Improve Maternal Health:**
   - Reduce by three quarters the maternal mortality ratio.

6. **Combat HIV/AIDS, Malaria and Other Diseases:**
- Halt and begin to reverse the spread of HIV/AIDS. Halt and begin to reverse the incidence of malaria and other major diseases.

7. **Ensure Environmental Sustainability:**
- Integrate the principles of sustainable development into country policies and programmes; reverse loss of environmental resources.
- Reduce by half the proportion of people without sustainable access to safe drinking water.
- Achieve significant improvement in lives of at least 100 million slum dwellers, by 2020.

8. **Develop a Global Partnership for Development:**
- Develop further an open trading and financial system that is rule-based, predictable and non-discriminatory, and includes a commitment to good governance, development and poverty reduction—nationally and internationally.
- Address the least developed countries' special needs. This includes tariff- and quota-free access for their exports, enhanced debt relief for heavily indebted poor countries, cancellation of official bilateral debt, and more generous official development assistance for countries committed to poverty reduction.
- Address the special needs of landlocked and small-island developing States.
- Deal comprehensively with debt problems of developing countries through national and international measures to make debt sustainable in the long-term.
- In cooperation with the developing countries, develop decent and productive work for youth.
- In cooperation with pharmaceutical companies, provide access to affordable essential drugs in developing countries.
- In cooperation with the private sector, make available the benefits of new technologies, especially information and communications technologies.

1.6.2 United Nations Environment Programme (UNEP): UNEP, established in 1972 after the Rio Earth Summit, is the principal entity within the UN System to assist the developing countries in building scientific and technical capacity, fostering partnership and knowledge development to promote environment for sustainable development. Based in Nairobi (Kenya), UNEP activities range from assessment of environmental trends—especially early warning systems to deal with environmental disasters and emergencies—to the promotion of environmental science and information.

One of the main responsibilities of UNEP is to keep under review the world environmental situation and ensure that emerging environmental problems of wide international significance are prioritized and receive appropriate and adequate consideration by the governments.

1.6.3 International Union for Conservation of Nature (IUCN): IUCN is a unique global organization which started working in 1948. It is one of the handfuls of international organizations where governments and non-governmental bodies work together as parties. By virtue of being a member of IUCN, a country has access to the largest network of specialists in the field of conservation. IUCN is having an observer status at the United Nations and advises governments on matters related to conservation, integrity and diversity of nature and also ensures that any use of natural resources is equitable and ecologically sustainable.

India has a long relationship with IUCN. The Government of India was the first country in South Asia to join IUCN as a state member in 1969. It was also the first country in the region to host the General Assembly of IUCN in 1969.

1.6.5 U.N. Framework Convention on Climate Change (UNFCCC): Recognizing the problem and the need for international co-operation, the international community in 1992 established the UNFCCC which accepted the principle of *common but differentiated responsibility* between the industrialized countries and the developing countries.

The objective of UNFCCC is to achieve stabilization of greenhouse gas concentrations in the atmosphere at a level that would prevent dangerous anthropogenic interference with the climate system. The Convention enjoins Parties to communicate information about the implementation of the Convention, taking into account the common but differentiated responsibilities and respective capabilities and their specific regional and national development priorities, objectives and circumstances. Article 12 of the Convention relates to the communication of information pertaining to implementation in accordance with Article 4(1) of the Convention, whereby each Party is required to communicate to the Conference of the Parties. The elements of information in the communication are provided in the context of national circumstances and include:

1. A national inventory of anthropogenic emissions by sources and removals by sinks of all greenhouse gases not controlled by the Montreal Protocol, to the extent its capacities permit, using comparable methodologies to be promoted and agreed upon by the Conference of Parties;
2. A general description of steps taken to implement the Convention including an assessment of impacts, vulnerability due to climate change and associated adaptation needs; and
3. Any other information that the Party considers relevant to the achievement of the objective of the Convention and suitable for inclusion in its communication, including, if feasible, material relevant for calculations of global emission trends.

To sum up, sustainable development is a pattern of social and structured economic transformations (i.e. development) which optimizes the economic and social benefits available at present, without jeopardizing the likely potential for similar benefits in the future. The primary goal of sustainable development is to achieve a reasonable and equitable distribution level of economic well being that can be perpetuated continually for the future human generations.

Sustainability is about meeting the needs of the present without compromising the ability of future generations to meet their needs. It is about preserving the environment and biodiversity for future generations, and about being cautious with our natural resources and climate. But sustainability is also about guaranteeing human rights and a life in dignity, free from want and poverty for all people living today.

Developing countries have options other than to *grow dirty and clean up later*. Much that is useful can be done now. Clean air and water and solid waste management are basic needs, and many environmental policies enhance productivity and poverty alleviation. So while poor countries should focus on meeting basic needs and expanding opportunities for growth, they need not do so at the expense of unsustainable environmental degradation. Further, environmental performance does not automatically improve with income, so policy action is needed anyway. Finally, it may be impossible or prohibitively expensive to *clean up later*, either because of the irreversibility of environmental damages like the loss of biodiversity, or because "lock-in" will make subsequent shifts to more environmentally benign structures and processes extremely costly.

2

History and Case Studies of Disasters

Disaster management is one of the front running themes among contemporary global issues. Disaster control and mitigation measures have been gradually attaining a level where it has been extensively practised. Disaster management has assumed great importance of late, due to the occurance of natural as well as man-made disasters. Not only the structual and physical measures, even the awareness about disasters in India is also minimal. Therefore, much more needs to be done to create better awareness on disaster management that could lessen the degree of damages. This could lead to a quicker and more honest efforts to respond to a disaster situation than what exists currently. Towards achieving this, the corporates, NGOs, government agencies and community as a whole have greater responsibilities. This research is a small effort in this direction.

National Disaster Management Act, 2005, defines disaster as "a catastrophe, mishap, calamity or grave occurrence affecting any area, arising from natural or man-made causes, or by accident or negligence which results in a substantial loss of life or human suffering or damage to, and destruction of, property, or damage to, or degradation of, environment, and such a nature or magnitude as beyond the coping capacity of the community of the affected area".

Disasters have adversely affected humans since the dawn of our existence. In response, individuals and societies alike have made many attempts to decrease their exposure to the consequences of these disasters, developing measures to address the initial impact, as well as post-disaster response and recovery needs. Disaster is a sudden, calamitous event bringing great damage, loss, and destruction and devastation to life and property. The damage caused by disasters is immeasurable and varies with the geographical location, climate and the type of the earth surface/degree of vulnerability.

2.1 History of Disasters

The history of disaster management practices can be found as early as 3,200 B.C. In what is now a modern day Iraq, lived a social group named as the Asipu. When community members faced a difficult decision, especially one involving risk or danger, they could appeal to the Asipu for advice. The Asipu, using a process which is similar to modern day hazard risk management, would first analyze the problem at hand, then propose several alternatives, and finally give possible outcomes for each alternative. Today, this methodology is referred to as 'decision analysis', and it is a key to any comprehensive risk management endeavour.

Early history is also marked by incidents of organized emergency response. For example, when in AD 79, the volcano Vesuvius began erupting two towns in its shadow, namely Herculaneum and Pompeii faced an impending catastrophe. Herculaneum, which was at the foot of the volcano and directly in the path of its lava flow, almost buried all the people immediately; the majority of Pompeii's population survived. This was because the citizens of Pompeii had several hours before the volcano covered their city in ash, and evidence suggests that city's leaders organized a mass evacuation which could save thousands of lives there. Modern form of disaster management is just an evolutionary stage of preceding plan policy and strategies which help the human beings in having a better preparedness against the disaster.

2.2 Orissa Super Cyclone, October 29-30, 1999

In India, the super cyclone of October 29-30, 1999, was undoubtedly the most intense one. It had some unique features such as, rapid intensification, small radius of eye-wall confining the large surge close to the point of landfall and relatively long life after the landfall. Climatologically, there is a high frequency of dissipation of cyclones in October because of strong easterly winds aloft. Cyclone genesis usually terminates at the stage of marginal cyclones.

The official death toll was 9,893, but there were difficulties in making accurate estimates and local people insist that the final death count was much higher. Almost all the villages along the sea in Ersama block of Jagatsinghpur district were washed away and no one knows how many people lost their lives. The cyclone also destroyed the livelihood of the coastal state's farming community—it saturated more than 1 million hectares of cropland under salty water and killed some 4,06,000 livestock (Disaster Relief Report). Millions of people who eked out their living on the land were left homeless and without a means to survive.

The cyclone struck just three weeks before the harvest and almost all the plantations in this mostly agricultural community—paddy fields, sugar cane, and vegetable crops—were destroyed. About 11 million people, nearly a third of the state's population of 35 million, were estimated by the UN agencies to be directly affected, having lost their shelter, crops, cattle and livelihoods.

This resulted in 4 different types of impacts:

- Physical destruction,
- Saline inundation,
- Flooding, and
- Psychological trauma.

Tables 2.1 to 2.4 show the extent of damage done by the Orissa Super Cyclone, 1999.

Other sectors which were badly affected by Orissa Super Cyclone 1999 (OSC) are as under:

- **Agricultural Sector:** OSC completely devastated the agricultural base and logistics in the affected areas of Odisha. In coastal belts, due to high tidal waves, the standing crops were damaged affecting 15 lakh families and the worst affected districts included Jagatsinghpur, Cuttack, Kendrapada and Puri.

- **Loss of Livestock:** More than 0.4 million cattle were killed by the super cyclone. Cattle deaths were reported highest in Jagatsinghpur, Kendrapara, Cuttack, Khudra and Puri.

Table 2.1: Districts with Maximum Human Casualties
(Orissa Super Cyclone, 1999)

District	Human Casualties	Population Affected		
		Total	Rural	Urban
Jagatsinghpur	8,119	13,62,760	15,99,295	64,117
Cuttack	471	24,17,048	18,47,923	5,69125
Kendrapara	469	13,70,000	13,03,200	75,800
Puri	303	15,63,000	13,70,000	1,93,000

Source: Gupta, M.C, "Orissa Super Cyclone, 1999", National Centre for Disaster Management.

Table 2.2: Districts with Maximum Washed Away Houses
(Orissa Super Cyclone, 1999)

Name of District	Fully washed away houses
Jagatsinghpur	1,2124
Balasore	11,483
Kendrapara	276
Mayurbhanj	262

Source: As in Table 2.1.

Table 2.3: Districts with Damaged Boats and Nets
(Orissa Super Cyclone, 1999)

District	Boats	Nets
Chilika Lake	7,560	11,599
Jagatsinghpur	6,988	16,271
Kendrapara	6,354	8,905
Puri	3,181	7,945

Source: As in Table 2.1.

Table 2.4: Districts with Maximum Number of School Damaged
(Orissa Super Cyclone, 1999)

Name of Districts	Number of schools damaged	
	Primary schools	High schools
Jajpur	2,115	208
Ganjam	1,972	315
Cuttack	1,617	424
Balasore	1,288	152
Jagatsinghpur	1,111	275

Source: As in Table 2.1.

- **Infrastructure:** Complete collapse of communication networks and the surface communication was hampered due to damage to the road and rail network. Water supply system and irrigation infrastructure was adversely affected. Drinking water sources were either destroyed or contaminated by strewn carcasses which further compounded the problems of already vulnerable state of the populace.
- **Research Institutes:** The super cyclone severely affected research institutes and facilities in the coastal belt of the state.

2.3 Community-based Disaster Preparedness (CBDP)

As a result of the Pilot Orissa Disaster Management Project, a local disaster management (preparedness and mitigation) system was installed in 10 blocks from the block level to the GP to the village levels. Increased level of alertness, especially with case stories of successful disaster preparedness activities in the June 2001 floods and November 2002 cyclone threat, was witnessed for the replication of the preparedness and mitigation activities in other blocks within the coverage districts of the project and for the other districts in Odisha.

Figure 2.1 presents a disaster preparedness-cum-social responsibility model.

Figure 2.1: Disaster Preparedness-cum-Social Responsibility Model

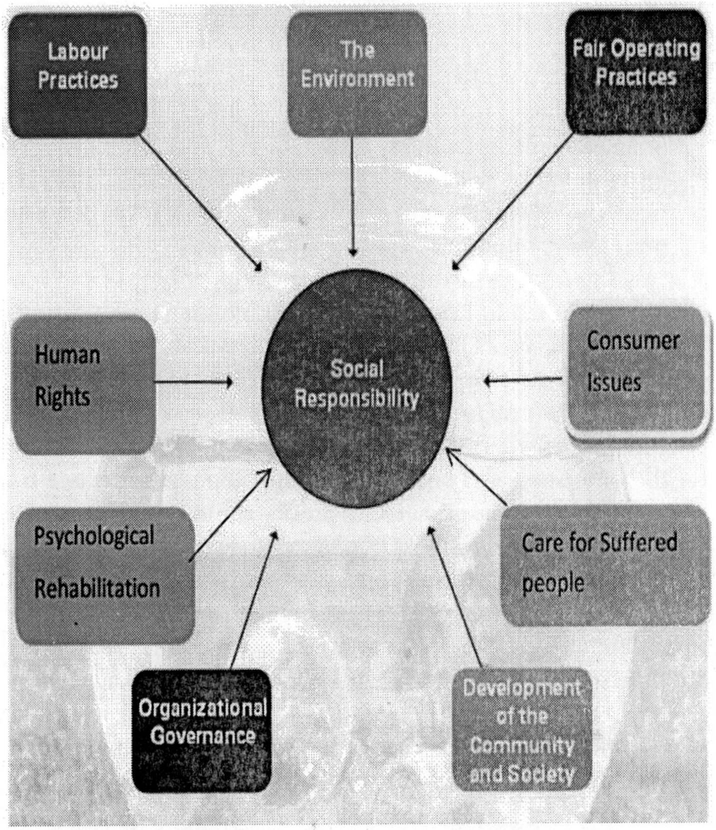

2.4 Tenth Five Year Plan (2002-07) and Orissa Super Cyclone

Tenth Plan—prepared in the backdrop of Orissa super cyclone—for the first time recognized disaster management as a development issue. The Plan devoted a separate chapter to disaster management and made a number of important prescriptions to mainstream disaster-risk reduction in the process of development.

The Plan prescriptions on disaster management could

broadly be divided into three categories:

1. Policy guidelines at the macro level that would inform and guide the preparation and implementation of development plans across sectors.
2. Operational guidelines for integrating disaster management practices into development plans and programmes.
3. Specific developmental schemes for prevention and mitigation of disasters.

At the macro level, the Plan emphasized that "while hazards, both natural or otherwise, are inevitable, the disasters that follow need not be so and the society can be prepared to cope with them effectively whenever they occur" and called for a "multi-pronged strategy for total risk management, comprising prevention, preparedness, response and recovery, on the one hand, and for initiating development efforts aimed towards risk reduction and mitigation, on the other". It stated that only then the country could look forward to sustainable development.

At the operational level, the Plan made a number of important suggestions as summarised below:

1. Streamlining institutional arrangements for disaster response by an integrated approach involving civilian and military resources.
2. Setting up a modern permanent national command centre or operations room with communications and data links to all State capitals.
3. Establishing a quick response team, particularly for search and rescue operations.
4. Developing standard operating system for dealing with humanitarian and relief assistance from non-governmental sources.
5. Formulating a unified legislation for dealing with all types of disasters.
6. Building disaster prevention and preparedness in development planning by introducing a rigorous process of

vulnerability analysis and risk assessment.

7. Establishing a disaster knowledge network for the use of disaster managers, decision-makers, community etc.
8. Developing a nation-wide culture of prevention by introducing disaster management in the school curriculum, including relevant aspects of disaster management in professional courses, enhancing the capacity of disaster managers by better training facilities and creating massive awareness at all levels.
9. Encouraging community level initiatives for disaster preparedness by involving people at the grassroots, particularly those who are more vulnerable, for better preparedness and response.
10. Developing appropriate zonal regulations, design standards, building codes and performance specifications for safe construction.
11. Inclusion of disaster mitigation analysis in all development schemes in vulnerable areas through which the feasibility of a project is assessed with respect to vulnerability of the area.
12. The Plan felt that planned expenditure on disaster mitigation and prevention measures was required in addition to the Calamity Relief Fund (CRF).

Thus, the Plan set into motion the process of shift in focus from response-centric disaster management—covering rescue, relief, rehabilitation and reconstruction—to laying greater emphasis on the other elements of disaster management cycle, viz. prevention, mitigation and preparedness as a means to avert or soften the impact of future emergencies.

2.5 Floods in Bangladesh

2.5.1 Understanding Floods: Technically, floods are any rise in water level in a stream and not only the ones that lead to an overflow of the normal confines of a stream, and are part of the river flows which constitute the total water resources. River flows causing flooding pose hazard and could result in disaster if not managed propoerly. According to Third World Water Forum, Kyoto, Japan, "floods have to be viewed from the socio-economic constructs of any region of the country and at the same time the potential of flood plains for socio-economic development cannot be overemphasised". The weaker sections of the society who are forced to occupy such hazardous areas are the most vulnerable. The continued increase in the economic utilization of the flood plains calls for the communities to be aware, that development activities in the flood plains increase the risk of flooding and there is a need to take adequate protective measures that are commensurate with the socio-economic benefits.

Among all the natural disasters afflicting the country frequently, river floods are the most devastating. River floods constitute one of the major natural hazards faced by India. Floods have been causing severe damage to man's interest in flood plains throughout the history and he has been constantly attemting to reduce the damages through the use of various types of flood prevention techniques. India is the worst flood hit country in the world after Bangladesh. India accounted for one-fifth of global death counts due to floods.

Human activity, however, tends to concentrate in flood livable areas which are often convenient and attractive locations for settlement and other economic endeavours resulting in greater flood damages. If losses due to floods in terms of lives and property are to be minimised, the solution is not merely to provide relief measures, but also to undertake necessary measures encompassing a wide range of activities like rehabilitation, resettlement and recovery of the victims. Therefore, the present study, highlights the same and similar issues in case of Bihar state also.

2.5.2 Background: Retrospective data suggests that Bangladesh is one of the most disaster prone countries in the world and it is not that Bangladesh alone suffers from these events; other countries also suffer from disaster vulnerabilities. But the frequency and vulnerabilities are more in Bangladesh.

Bangladesh is a low-lying deltaic country in South Asia formed by the Ganges, the Brahmaputra and the Meghna Rivers, where 145 million population is concentrated in 1,47,570 sq. km. territory and more than 230 waterways and tributaries have made this country a land of rivers. Among all the natural disasters afflicting the society, river floods are the most devastating, which cause maximum damages of life and property in countries like Bangladesh. The causes of floods in Bangladesh are mainly due to the peculiarity of tributaries and confluence of big rivers like Meghna, Jamuna, Ganga and Brahmaputra.

The flood plains which occupy 80 percent of the Bangladesh's geographical area have diversified characteristics and are affected by flash floods, river floods and rain floods. The maximum volume of water (749 million m3) is carried by the Brahmaputra river alone. The country has the world's longest unbroken sandy beach of 120 km. sloping gently down to the blue waters of the Bay of Bengal. Annually, an average of 15 percent of the total geographical area is inundated by floods. Floods in 1988, 1998, and 2004 caused huge damage of resources amounting to US$ 10.6 billion (Table 2.5).

Table 2.5: Average Annual Loss Due to Floods

	Item	Loss
1.	Land area affected	7.56 million hectare
2.	Population affected	32.03 million
3.	Human lives lost	3,000
4.	Livestock lost	3,00,000
5.	Houses damaged	11,683
6.	Crops damaged	Tk 700 crore
7.	Public utility damaged	Tk 1,000 crore
8.	Population displaced	20.10 million
9.	Population affected by communicable diseases	1.50 lakh

Source: Official Data of Ministry of Agriculture, RoB (2007).

There is a growing feeling that the incidence and intensity of floods has grown alarmingly over the years. A major cause is the increased encroachment of flood plains because of the new developments and population pressure. The damage caused by floods can at best be minimized and not altogether eliminated or in other words, there can really be no such things as fool proof protection or absolute flood control for all the magnitudes of floods. The concept of flood management, therefore, aims for such planned measures which ensure profitable and economic utilization of the flood plains. There are some other reasons for floods also. They are as follows:

1. Inadequate capacity of banks of rivers to resist high flow.
2. River bank erosion and silting of river beds.
3. Landslides leading to obstruction of flow and change in the river course.
4. Flow retardation due to tidal and bank water effects.
5. Synchronizations of flood in the main and tributary river course.
6. Poor natural drainage system.
7. Cyclone and heavy torrential rains.

The 1998 floods in Bangladesh had a substantial impact on the health of community. Diarrhoea was associated with socio-economic status, water handling and household sanitation. The flood had ravaged approximately 60 percent of the land and affected over 30 million people.

The inhabitants of flood plains of Bangladesh's main river received little government support to cope with floods and developed their own strategies to combat it. Major flood projection works were not possible in these areas where floods and erosion affected many vulnerable people. It was found that total losses and dislocations to their lives were substantial.

Vulnerability is one among many concepts associated with disasters. Different concepts of vulnerability are given by the social scientists. The potential for attributes of a system to respond adversely to the concurrence of hazardous events is vulnerability. A characteristic of individuals and groups of people who inhabit a given natural, social and economic space, within which they are differentiated according to their varying positions in the society into more or less vulnerable individuals or groups is also considered of it.

Vulnerability is the propensity of an endangered event due to any kind of natural hazard to suffer different degrees of loss or amount of damage depending upon its particular social, economic, cultural and political weaknesses (Table 2.6).

Table 2.6: Major Floods in Bangladesh and Causalities

Year	Places mostly affected	Causalities
1644	Chittagong	Not Available
1648	Sirajganja	,,
1871	Rajsahi	,,
1885	Rajshai, Dinajpur	,,
1892	Rajshai, Dinajpur	,,
1906-7	East Bengal	,,
1918	Jessore	,,
1922	Comilla	,,
1955	Chandpur	,,
1974	Jehnaidah	30,000
1975	Joipurhat	
1984	Bogra	
1987	Dhaka	1,657
1988	Syllhet	2,379
1998	Joypurhat, Naogaon, Chapainnawabganj	1,050
1999	Maulavibazar	
2001	Eastern & Northern Bangladesh	
2004	Dhaka and central Bangladesh	481
2007	Comilla, Dhaka, Sylhet	226

Source: Field Data and UNICEF Report, 2007.

2.5.3 Morbidity (Illness, Injuries and disease): Injuries, illness and diseases are associated with the flood; specifically the injuries may happen as individuals attempt to remove themselves, their family, or valued possessions from danger.

There is potential for injuries when people return to their homes and businesses and begin with the clean-up operation, vacating debris and all. Both fatal and non-fatal diseases are mostly routinely reported in Bangladesh floods. In flood conditions, there is potential cholera like epidemic, water borne and vector borne diseases like diarrhoea, skin diseases, infection, snake bites, dysentery, pneumonia and respiratory tract infection etc. Table 2.7 shows that in the Bangladesh 2007 flood, out of the diarrhoea cases reported, 30,000 seemed to be more serious and acute. Though minimal cases relating to snake bites were reported but these cases were more fatal. Vector borne diseases like malaria and many infections are also transmitted by mosquitoes, which breed in stagnant waters.

The World Health Organization (WHO) recognizes that the mental health consequences are also not fully addressed. It is generally accepted that natural disasters like earthquake, flood, cyclone etc. create an aura of mental trauma, anxiety, and depression among the people mostly living in developing countries, where capacity to take care of these problems is not possible.

Table 2.7: Types of Diseases, Illness and Injuries in 2007 Bangladesh Flood

Diarrhoea	1,75,000
Acute Diarrhoea	30,000
RTI	38,000
Eye infection	2,304
Snake bites	132
Injuries	2,229
Pneumonia	900
Water borne diseases	4,133
Cholera and epidemic	1,200
Other reasons	36,392
Respiratory diseases	1,190

Source: UNICEF Report, 2007.

2.5.4 Mortality: The most readily identified flood deaths are those that occur from drowning or trauma, such as being hit by objects in fast flowing water currents. The number of such cases is determined by the characteristics of flood. The risk of deaths most obviously increase during the flood period. In a study of 2004 Bangladesh floods, it was found that most of deaths occurred due to water borne diseases like diarrhoea and dysentery.

The 2004 floods of Bangladesh appeared to be more furious in nature as several morbidity cases were reported. Along with pneumonia, diarrhoea, snake bites, and water borne diseases, vector borne diseases like dengue fever were also reported from almost all parts of the country during the entire flooding stint. A number of dengue cases were reported between July 12 and August 25, 2004. The highest number of dengue cases was found on July 18 and the lowest on August 14, 2004.

Unhygienic and improper sanitary conditions of the locality cause the unprecedented situation. The relation between flooding and dengue, malaria or such other vector borne diseases are complex. Many important infections are transmitted by mosquitoes. Storage of stagnant water due to the blocking of drains and water logging, especially in urban settings, can also be associated with increase in transmission of Dengue like vector borne diseases.

In the 2004 flood, a total of 2,316 cases of dengue fever with 5 deaths were reported till August 21, 2004. The general trend was that as flood water recedes, the prevalence of diarrhoea diseases increases. The number of reported cases of diarrhoea however decreased, with a peak observed during the first week of August 2004.

2.5.5 Migration/Displacement: A major part of demographic transition is migration and displacement. Bangladesh floods lead to temporary migration or so called displacement for time being as the devastating torrential and flush floods wash away a number of habitats, houses, home stead lands, agricultural and semi-urban centres. Poor people lose their live sustenance during the flooding. Not only that, in order to save themselves lives from the devastating floods, people tend to displace and evacuate to safer place in remote rural highlands and other urban centres and take shelters. Table 2.8 depicts the number of displacement in the floods of 2007.

Table 2.8: Displacement during the 2007 Floods

Number of people affected	6,35,946
Number of people displaced from their homes	33,500
Number of shelter houses provided	145

Source: UNICEF Report, 2007.

3

Pipeline Disasters and Oil Spills

A disaster is a serious disruption of the functioning of a community or a society involving widespread human, material, economic or environmental losses and impacts which exceeds the ability of the affected community or society to cope up using its own resources.

Contemporary disasters are seen as the consequence of an inappropriately managed risk.

These risks are the product of a combination of both hazard/s and vulnerability. Hazards that strike in areas with low vulnerability will never become disasters, as is the case in uninhabited regions. Losses due to natural hazards are 20 times greater (as a percentage of GDP) in developing countries than in industrialized countries.

India has been traditionally vulnerable to natural disasters on account of its unique geo-climatic conditions and about 30 million people are affected by disasters every year. The loss in terms of private, community and public assets has been astronomical.

3.1 Pipeline Transportation

Accidents in oil and pipeline industries constitute a major threat to property and population because of the magnitude of the hazard. The pipeline transportation is quite vast in India, supplying oil and natural gas; so there are innumerable causes that lead to pipeline disasters of major or minor nature.

The main aim of the study is to examine and know the reason for the probable cause of Bhadbhut pipeline disaster in 2013.

The purpose of the report is to analyze and know the various factors which contributed to the rate of occurrence of incidents in the transmission of oil and gas pipelines in Gujarat, the role of the administration and stakeholders and apart from these, what were the factors considered while laying these pipelines in Gujarat. In some cases, delayed controlling of oil spill led to irreparable damages.

Moreover, what is the response time to control the oil pipeline spillage, accordingly decides the damage done. Thus, an immediate and effective response system against an oil spill can reduce the incurred costs significantly.

At the global level, there has been considerable concern over natural disasters. Pipeline transport is the transportation of goods through pipes. Pipeline transport for oil and natural gas seems to be the most preferred, convenient, economical and reliable method.

Liquids and gases are transported in pipelines and any chemically stable substance can be sent through a pipeline. These pipelines used for the transportation of oil or gas have been considered as relatively safe tools for many years.

However, pipelines have caused huge number of accidents so far and the failure of oil pipelines could be the result due to various causes including technical reasons or external impacts as well as due to human errors. These causes vary depending on the type of pipeline network, local regulation, and ecological circumstances.

A wide range of regulatory regimes have addressed safety requirements for prevention of accidents based on potential risk. Pipeline accidents affect the daily lives of people in most parts of the world. Modern day life is based on structures in which energy fulfils a prevailing role.

India had about 125 million metric tonne of proven oil reserves which is the second largest amount in the Asia-Pacific region behind China.

As per the Ministry of Petroleum, India has 1,437 billion cubic metres (50.7×10^{12} cu ft) of confirmed natural gas reserves. Currently, India has over 13,428 km. long gas pipeline network, with a transmission capacity of 355 MMSCMD.

3.2 Oil Spills

Oil spill is a release of liquid petroleum hydrocarbon into the environment, especially marine areas, due to human activity and is a form of water and air pollution.

The term oil spill is usually applied to marine oil spills, where oil is released into the ocean or coastal waters, but spills may also occur on land.

Oil spills may be due to the release of crude oil from tankers, pipelines, offshore platforms, drilling rigs and wells, as well as spills of refined petroleum products (such as gasoline, diesel) and their by-products, heavier fuels used by large ships such as bunker fuel, or the spill of any oily refuse or waste oil.

A huge mass of India's natural gas production comes from western offshore region like Gujarat. Gujarat state is located in the western part of India and its oil and gas pipeline network is the largest compared to any other state in India.

Presently, the length of oil and gas pipeline is over 22,000 km. in the state and further extension of pipeline network is underway.

Through this pipeline network, 27.3 million cubic feet of gas and 6,64,582 barrels of oil per day is transported. The oil fields of Cambay region in Gujarat on land and gas from Hazira fields in south offshore Gujarat are some of the major resources. The city gas grid network has been initiated recently and gradually will cover all the cities.

Nevertheless, over one billion gallon of oil and gas are still spilled every year. For instance, the oil pipeline spillage in August 2013 near Bhadbhut village in Gujarat which was caused by a leak within 30 inches gas pipeline led to loss of gallons of oil spilled into the Narmada River.

The problems due to present disaster occurred after the explosion of oil and gas pipeline in Bhadbhut village in Bharuch which led to oil spill in the entire north section of the Narmada River and after the disaster, not a single operator accepted the responsibility of the disaster.

Each operator held the other one responsible for the disaster.

On the other side, the administration took a strict action against them by sending them notices and taking serious steps in reviewing the entire pipeline laying norms and suggesting a policy issue problem after the disaster incident.

An oil spill was also reported in the coastal areas of Gulf of Khambhat following a rupture in Oil and Natural Gas Corporation's (ONGC) pipeline near Bhadbhut village of the Bharuch district.

The rupture in ONGC's pipeline was brought to notice by the Gujarat Pollution Control Board and they were asked to take immediate action to control the situation.

Following that, the Indian Coast Guard was informed about the spillage, which undertook an aerial survey of the affected area using its advanced light helicopter 'Dhruv'.

ICG, in its aerial survey, found that the spill had spread in the area of 2-3 nautical miles in the sea in the Gulf of Khambhat. Oil spill dispersant (OSD) was not used because the situation of water was confined and shallow.

The district administration was allowed to deploy the mechanized boat so that they can create a churn around the oil-spilled region, which can break the oil into droplets.

Oil spill dispersant is a mixture of surfactants and solvents that break up an oil spill into droplets.

By breaking it up, microbes and the environment can more easily biodegrade the oil. The effect of the spill on the environment is yet to be ascertained (Singh, 2013).

Rupture in a pipeline led to a scare in villages along the Narmada River in Bhadbhut area of Bharuch district. The blast was caused by a rupture in the pipeline of Gas Authority of India Limited (GAIL). The villagers heard a deafening sound in the river that led to a scare among people.

Soon residents of Bhuva, Kansva, Navetha, Shankhwad, Mothia, Manad, Mahegam and other villages called the residents of Bhadbhut to enquire about the incident.

Actually, there was smoke and smell of gas all around and the villagers realized that one of gas pipelines passing under the river had been ruptured. The rupture had occurred in the 30-inch diameter gas pipeline of GAIL.

There are five gas pipelines passing across the river out of which four were operational. As for precaution, the gas supply from all the pipelines was stopped immediately and was resumed later only after ensuring their safety.

As explained by the villagers to the local administration, heavy floods led to soil erosion and the consequent rupture.

3.3 Nigerian Dilemma

Since its discovery in Nigeria in 1956, crude oil has been a source of mixed blessing for the country. It is believed to have generated enormous wealth, but has also claimed great many lives.

The impact of oil on security in Nigeria has largely focused on internal conflicts rather than on how disasters associated with oil pipeline vandalisation have impacted the human security in terms of causing injuries and deaths, destroying livelihoods and fracturing families.

Women and children as those who are the hardest hit and raises questions as to why the poor are the most vulnerable in oil pipeline disasters in the country.

Stress has been laid on adoption of a comprehensive and integrated framework of disaster management that will ensure prompt response to key early warning signs.

Over the past few decades, four distinct and largely independent research and policy communities in the areas of disaster risk reduction, climate change adaptation, environmental management and poverty reduction have been actively engaged in reducing the socio-economic vulnerability to natural hazards.

However, despite the significant efforts of these communities, the vulnerability of many individuals and communities to natural hazards continues to increase considerably.

In particular, it is the hydro-meteorological hazards that affect an increasing number of people and cause increasingly large economic losses.

Arising from the realisation that these four communities have been largely working in isolation and enjoyed only limited success in reducing vulnerability, there is an emerging perceived need to strengthen significant collaboration and to facilitate learning and information exchange between them.

This calls for developing models for understanding key communalities and differences between the climate change adaptation and disaster risk reduction communities, and proposes three exercises that would help to structure a multi-community dialogue and learning process.

All models seek to describe the key physical and chemical processes that transport and weather the oil on and in the sea. Insights into the mechanisms of these processes and the availability of algorithms for describing and predicting process rates are being explored.

Advances are noted in the areas of advection, spreading, evaporation, dispersion, emulsification and interactions with ice and shorelines.

Knowledge of the relationship between oil properties and oil weathering and fate and the development of models for the evaluation of oil spill response strategies are being summarized. Specific models are used as examples where ever appropriate to devise appropriate strategies.

Many factors play an important role in the effectiveness of technology. Effective communication networks established prior to a crisis can help to overcome distrust within the community as well as towards government or other outside agencies.

Disaster is a catastrophic situation in which the normal pattern of life or eco-system has been disrupted and extra-ordinary emergency interventions are required to save and preserve lives and or the environment. When communities and disaster agencies understand how neighbourhoods, towns, and regions relate to each other, they can determine how easy or difficult it will be for individuals and families, communities and the districts to come together in an emergency, rely on each other, and communicate essential information to one another when needed. The most dynamic networks are built on trust and mutual exchange.

4

Climate Change and Disasters

India has always been subject to a large degree of climate variability. This is likely to be accentuated by climate change. Warming trends along the west coast, in Central India, the interior peninsula, and North-Eastern India and cooling trends in North-West India and parts of South India have been observed. Coasts are projected to be exposed to increase in risk including coastal erosion due to climate change and sea level rise. Geological Survey of India (GSI) projects that the glaciers of Himalaya are receding at varying rates in different regions. India is conscious of these challenges and has implemented several major programmes addressing the climate variability concerns. These include the following.

- Cyclone warning and protection.
- Coastal protection.
- Floods and drought control and relief.
- Major and minor irrigation projects.
- Control of malaria.
- Food security measures.
- Research on drought resistant crops.

4.1 Climate Change and Human Settlements

Climate change causes vulnerability of human settlements which is related to extreme weather events and such gradual changes in the climate exceed the adaptive capacity of human systems. Climate change adds to the existing stress on the sustainability of human settlements and society. Non-climate sources of change like rapid urbanisation are often the main source of stress.

Human settlements are accustomed to variability in

environmental conditions and are resilient to normal variations. Vulnerabilities arise out of experiences that are beyond the normal experience and due to limited adaptive capacity. Climate change is likely to affect infrastructure related to water, sanitation, energy, transportation, healthcare, fire services and other forms of emergency measures.

The urban-heat island effects could get exacerbated due to increase in baseline temperatures, affecting climatic comfort of the urban populations and may consequently lead to additional costs in climate control.

Sustainable development concerns in the sense of enhancement of human well-being broadly conceived, are a recurring theme in India's development philosophy. The present day consensus reflects three foundational aspirations. *First, that human beings should be able to enjoy a decent quality of life; second, that humanity should become capable of respecting the finiteness of the biosphere; and third, that neither the aspiration for the good life, nor the recognition of biophysical limits should preclude the search for greater justice in the world.*

Sustainable habitat would mean achieving a balance between the economic and social development of human habitats together with the protection of the environment, equity in employment, shelter, basic services, social infrastructure and transportation. In the context of rapid urbanisation, there is wide awareness about the need to minimize the environmental costs of urbanisation. Environmental damage and depletion of non-renewable resources need to be addressed to meet present and future challenges such as climate change and associated vulnerabilities.

4.2 Climate Change and Water Supply

Climate change could affect water supply systems in any number of ways. It can affect the water demand for drinking and cooling systems. Where climate change leads to failure of small local water sources such as wells, it could lead to greater demand for regional water supplies. Changes in precipitation patterns could lead to reduction in water availability and fall in water tables. In coastal areas, it could lead to saline intrusions in rivers and ground water. Loss of melt water could reduce river flows during critical times.

A change in water availability and supply also affects sewered sanitation and drainage systems. When water supplies reduce, sewerage systems also become vulnerable. Further, sewage treatment plants are vulnerable to floods or sea level rise, as these are often located near rivers or seas. Where the sewer outfalls are in the sea, sea level rise will affect the functioning of such systems. Storm water drainage systems could become frequently overloaded and cause flooding if heavy storms become more frequent due to climate change. The impact of inadequate drainage systems in cities like Mumbai is already being felt leading to flooding and huge economic losses. More frequent floods could also present a significant threat if these lead to contamination of flood waters with faecal material.

4.3 Occurrence of Droughts

One of the climate change risks to the Indian economy and its people, and therefore to sustainability of habitats, is the increased intensity, frequency and geographical coverage of drought. Its primary impact is on rural areas where agriculture, animal husbandry, and to a lesser extent forestry and fishing, are significantly impacted leading to cycles of seasonal and distress migration. Droughts lead to drinking water shortages and increase in food and biomass fuel prices that hurt the urban poor and middle classes and hence the city economy. It also leads to a number of other impacts like depressed demand for secondary goods and services because of depressed agricultural demand. Climate change related impacts may render subsistence agriculture uneconomical in parts of semi-arid central, western and southern India, fuelling migration of rural population to urban areas.

Climate change may increase droughts in semi-arid peninsular and western India, leading to further misery for the landless and small and marginal farmers, who are typically forced to migrate to cities. They form the most vulnerable groups in cities—having limited skills, education, capital and access to social networks that underpin much economic and social mobility in urban India. They are also more likely to live in illegal and unserviced settlements that are exposed to a wide range of environmental risks varying from flooding to fire and continual cycles of demolition and eviction by civil authorities. They are therefore, dual victims of existing natural hazards and emerging climate change—displaced from their original places of residence and occupations and challenged by urban risks at their new places of residence.

4.4 River Hydrology and Inland Flooding

The most serious regional impact of climate change could be changes in the river hydrology in the Indo-Gangetic plain and the Brahmaputra valley due to glacial melt and regression of the Himalayan glaciers. The sustainability of habitats in these regions would be at risk. Inter-basin transfers could be amongst the options that might need to be considered.

Climate change is also expected to increase the severity of flooding in many Indian river basins, especially of the Godavari and Mahanadi along the eastern coast. Floods are expected to increase in north-western India adjoining Pakistan, and in most coastal plains in spite of existing upstream dams and multi-purpose projects. Extreme precipitation is expected to show a substantial increase in some parts of the country. Sanitation infrastructure is the main determinant of the contamination of urban flood water with faecal material presenting a substantial threat of enteric diseases. In addition to flood hazards, more extreme rainfall events associated with climate change will also cause hazards from landslides in many urban centres in hilly regions.

Dealing with likely increase in risks associated with storm

water will require a significant revision of urban planning practices across cities in the flood affected regions. Flood and climate change mitigation and adaptation measures will have to be integrated into day-to-day urban development and service delivery systems. Due to higher precipitation in more intense events, there is a need to increase the margin of design safety to enhance the hydraulic capacity of the sections. This will require additional investments.

4.5 Cyclonic Storms, Storm Surges and Coastal Flooding

Yet another important climate change induced risk is that of cyclonic storms, storm surges and accompanying coastal inundation. The high concentration of population, especially on the eastern coast, has led to extremely high vulnerability in this region, leading to devastating loss of life and property. The 1999 Orissa super cyclone killed over 10,000 people, devastated buildings, lifeline infrastructure and economic assets across 10 coastal and six inland districts, which included a number of towns and cities due to a mixture of devastating storm surges, cyclonic winds and coastal flooding.

Cyclones and storm surges could have a devastating impact on large urban centres including the mega cities of Mumbai and Chennai, the million plus cities of Vishakapatnam and Surat as well as other cities like Bharuch, Bhavnagar and Jamnagar apart from causing critical bottlenecks in important ports such as Kandla. Increased migration to the coast, powered by huge investments in coastal infrastructure, settlements and enterprises could lead to a substantial rise in losses.

4.6 Sea Level Rise

Rising sea levels in coastal areas would increase the risk of flooding and increase the vulnerability of communities residing in these areas, especially the poor.

It is estimated that sea level could rise by 3.5 to 34.6 inches between 1990 and 2100, making coastal groundwater saltier, endangering wetlands, and inundating valuable land and coastal communities. The most vulnerable stretches along the western Indian coast are Khambhat and Kachchh in Gujarat, Mumbai and parts of the Konkan coast and south Kerala. The deltas of the Ganga, Krishna, Godavari, Cauvery and Mahanadi on the east coast may be threatened, along with irrigated land and a number of urban and other settlements that are situated in them. The loss of these important economic and cultural regions could have a considerable impact in some states.

There is also a need for better empirical data regarding storm surge and tidal gauge data for better approximation of the impact of climate change in coastal areas. Tools like micro-modelling could help in developing better understanding.

According to Inter-Governmental Panel on Climate Change (IPCC), Indian coasts will be exposed to increasing risks such as coastal erosion due to climate change and sea level rise. Sea water intrusion is likely to increase the habitat of brackish water fisheries but coastal inundation is likely to damage the aquaculture industry significantly. Changes in currents, water temperature, salinity, strength of upwelling and mixing layer thickness in the West Pacific and North Indian Oceans are expected due to climate change.

Sea level rise will exacerbate the already declining fish productivity in Asia. Flooding caused by sea level rise is expected to affect millions of additional people, with small islands and the crowded delta regions around large Asian rivers (such as the Ganges and Brahmaputra) facing the highest risk. The IPCC has identified four deltas along the coast of India that are particularly vulnerable: Ganges-Brahmaputra (extremely vulnerable), Godavari (highly vulnerable), and Krishna and Mahanadi (medium vulnerability). Stability of wetlands including mangroves and coral reefs around Asia is likely to be increasingly threatened.

4.7 India's Approach to Climate Change

The climate of India may be broadly described as tropical monsoonal type. India's climate is affected by two seasonal winds: the North-East monsoon and the South-West monsoon. The North-East monsoon, commonly known as the winter monsoon blows from land to sea, whereas the South-West monsoon, known as the summer monsoon blows from sea to land after crossing the Indian Ocean, the Arabian Sea, and the Bay of Bengal. The South-West monsoon brings most of the rainfall during a year in the country.

India is a large developing country with nearly 700 million rural population directly dependent on climate-sensitive sectors (agriculture, forests and fisheries) and natural resources (water, biodiversity, mangroves, coastal zones, grasslands) for subsistence and livelihoods. Further, the adaptive capacity of dry land farmers, forest dwellers, fisher folk and nomadic shepherds is very low.

Climate change may alter the distribution and quality of India's natural resources and adversely affect the livelihoods of its people. With an economy closely linked to its natural resource base and climatically-sensitive sectors, India may face a major threat because of the projected change in climate. With climate change, there would be increasing scarcity of water, reduction in yields of forest biomass, and increased risk to human health. The contribution of India to the cumulative global emissions is only 5 percent. Thus, historically and at present, India's share in the carbon stock in the atmosphere is relatively miniscule when compared to its distribution over the nation's population.

Climate change, resulting from anthropogenic emissions and increasing concentrations of a suite of gases called *green house gases* (GHGs)—mainly due to fossil fuel use, certain agricultural and industrial activities, and deforestation—has the potential, over the next few generations, to significantly alter global climatic conditions. This would result in large changes in ecosystems, leading to possibly catastrophic disruptions of livelihoods, economic activity, living conditions, and human health. On the other hand, the abatement of GHGs involves significant economic costs.

While climate change is a global environmental issue,

different countries bear different levels of responsibility for increases in atmospheric GHG concentrations. Further, the adverse impacts of climate change will fall disproportionately on those who have the least responsibility for causing the problem, in particular, developing countries including India.

Though India is a signatory to UNFCCC, it is not required to contain its GHG emissions. India's policies for sustainable development—by way of promoting energy efficiency and the use of renewable energy, changing the fuel mix to cleaner sources, energy pricing, pollution abatement, afforestation, mass transport—themselves result in a relatively GHG benign growth path.

4.8 National Action Plan on Climate Change (NAPCC)

As a part of voluntary actions to address climate change related concerns, India launched NAPCC on June 30, 2008. It stresses that maintaining a high growth rate is essential for increasing living standards of the vast majority of people of India and reducing their vulnerability of the impacts of climate change. Accordingly, it identifies measures that promote the objectives of sustainable development of India while also yielding to benefits for addressing climate change.

NAPCC estimated the per capita emissions of carbon dioxide in India at 1.02 metric ton per capita which is well below the world average of 4.25 and that of China at 3.60. The energy intensity of the economy is also demonstrated as having declined very considerably since the 1980s, to the present level comparing favourably with that of the least energy-intensive developed countries.

The NAPCC is co-ordinated by the Ministry of Environment, Forest and Climate Change (MoEFCC) and implemented through the nodal Ministries and is aimed at advancing relevant actions in specific sectors/areas.

NAPCC advocates a strategy that promotes, firstly, the adaptation to climate change and secondly, further enhancement of the ecological sustainability of India's development path.

Eight national missions—in the areas of solar energy, enhanced energy efficiency, sustainable agriculture, sustainable habitat, water, Himalayan ecosystem, increasing the forest cover, and strategic knowledge for climate change—form the core of NAPCC. These National Missions, which form the core of NAPCC, represent multi-pronged, long-term and integrated strategies for achieving key goals in the context of climate change. The focus is on promoting understanding of climate change, adaptation and mitigation, energy efficiency and natural resource conservation.

State Governments are also preparing—under the advice of the Central Government—State Action Plans aimed at creating institutional and programme-oriented capacities to address climate change. These, together with the National Missions, will enhance climate change-related actions in the public and private domains.

4.9 Eight National Missions of NAPCC

4.9.1 Jawaharlal Nehru National Solar Mission (JNNSM): The government launched the JNNSM in January 2010 with a target of 20,000 MW grid solar power (based on solar thermal power generating systems and solar photovoltaic [SPV] technologies), 2,000 MW of off-grid capacity by 2022. JNNSM will be implemented in three phases. The first phase will last three years (up to March 2013), the second till March 2017, and the third till March 2022.

A. Mission Objective: To establish India as a global leader in solar energy, by creating the policy conditions for its diffusion across the country as quickly as possible.

B. Mission Targets and Timeline: The Mission has adopted a three phase approach. The first phase (2010-2013) was designed to focus on capturing the low-hanging options in solar thermal; promoting off-grid systems to serve populations without access to commercial energy and modest capacity addition in grid-based systems. In the second (2013-2017) and third (2017-2022) phases, capacity will be aggressively ramped up to create conditions for scaled-up and competitive solar energy penetration in the country. To achieve this, the mission targets that by the year 2022 it will:

- Create an enabling policy framework for the deployment of 20,000 MW of solar power.
- Create favourable conditions for solar manufacturing capability, particularly solar thermal for indigenous production and market leadership.
- Promote programmes for off grid applications reaching 2,000 MW.
- Achieve 20 million sq. meters solar thermal collector area.
- Deploy 20 million solar lighting systems for rural areas.

4.9.2 National Mission for Enhanced Energy Efficiency (NMEEE): The objective of NMEEE is to achieve growth with ecological sustainability by devising cost-effective strategies for end-use demand side management. The Ministry of Power and Bureau of Energy Efficiency have been entrusted with the task of preparing the implementation plan for the NMEEE and up scaling efforts to create and sustain market for energy efficiency to unlock investment of around ₹ 74,000 crore. NMEEE is expected to achieve about 23 million tons oil-equivalent of fuel savings in coal, gas, and petroleum products by 2014-15, along with an expected avoided capacity addition of over 19,000 MW. The carbon dioxide emission reduction is estimated to be 98.55 million tons annually.

A. Mission Objective: To achieve growth with ecological sustainability by devising cost-effective and energy-efficient strategies for end-use demand side management.

B. Mission Targets and Timeline: To achieve its objective, the mission focuses on the following initiatives:

- **Perform Achieve and Trade (PAT):** A market-based mechanism to facilitate energy efficiency improvements in large energy intensive industries and facilities, by issuing energy saving certificates that can be traded.
- **Market Transformation for Energy Efficiency (MTEE):** Accelerating the shift to energy-efficient appliances and

equipments in designated sectors through innovative measures that make such products more affordable.

- **Energy Efficiency Financing Platform (EEFP):** Creating mechanisms to finance demand side management programmes in all sectors of the economy by capturing future energy savings.
- **Framework for Energy Efficient Economic Development (FEEED):** Developing fiscal instruments to promote energy efficiency.

By 2015, implementation of the mission is expected to deliver estimated fuel savings of about 23 million tonnes of oil-equivalent every year, along with avoided capacity addition of over 19,000 MW. The resultant annual reduction in carbon dioxide emissions is estimated to be around 98.55 million tonnes.

4.9.3 National Mission on Strategic Knowledge for Climate Change (NMSKCC): NMSKCC was launched with the following objective and targets:

A. Mission Objective: To identify the challenges and the responses to climate change through research and technology development and ensure funding of high quality and focused research into various aspects of climate change.

B. Mission Targets and Timeline: To achieve its objective, the mission targets are as follows:

- Form well-designed knowledge networks with a well-structured framework for harmonisation, interoperability, sharing and exchange of data of relevance to climate change and responses.
- Enhance the research capability in climate science.
- Position a technology watch system for key sectors related to economic development, likely to be affected by climate change.
- Leverage development of suitable technologies for adaptation and mitigation of climate change under various missions.
- Assist other agencies engaged in the implementation of the National Action Plan on Climate Change and supporting

the actions under the other Missions, as and if necessary.

4.9.4 National Mission for Sustaining Himalayan Ecosystem (NMSHE): The objective and targets of NMSHE include the following:

A. Mission Objective: To evolve management measures for sustaining and safeguarding the Himalayan glaciers and mountain ecosystem and attempt to address key issues, namely impacts of climate change on the Himalayan glaciers, biodiversity, wildlife conservation and livelihood of traditional knowledge societies.

B. Mission Targets and Timeline: To achieve its objective, the mission's selected targets are as under:

- Creation of a fund of ₹ 1,650 crore for developing capacities for sustaining Himalayan ecosystem.
- Establishment of a state-of-the-art National Centre for Himalayan Glaciology.
- Identification and networking of all knowledge institutions in the region which possess the institutional capacity for studies on Himalayan ecosystems.
- Establishment of about 10 new centres in existing institutions in areas of knowledge gaps complete with special mechanisms and tools to create knowledge capacity for sustaining Himalayan ecosystems.
- Annual status reports on the health of various sub-components of the Himalayan ecosystems and bi-annual advisories to the Himalayan Sustainable Development Forum through state councils for climate change in the Indian Himalayan states for actions for implementation.
- Standardisation of data collection systems for inter-operability and mapping of natural resource wealth systems.
- Identification and training of about 100 experts and specialists in areas relevant to sustaining the Himalayan ecosystem including about 25 glaciologists.
- Conduct 25 programmes on capacity building for linking innovations from traditional and modern knowledge systems.

- Establishment of an observational network for monitoring and forewarning of changes in the ecosystems of the Himalayan region.

4.9.5 National Water Mission (NWM): The objective and targets of NWM are as under:

A. Mission Objective: To conserve water, minimise wastage and ensure equitable distribution both across and within states through integrated water resources development and management.

B. Mission Targets and Timeline: To achieve its objective, the mission's targets are as follows:

- Development of comprehensive water database in public domain and assessment of impact of climate change on water resources.
- Promotion of citizen and state actions for water conservation, augmentation and preservation.
- Focused attention to vulnerable areas including over-exploited areas.
- Increase water use efficiency by 20 percent.
- Promotion of basin level integrated water resources management.

4.9.6 Green India Mission (GIM): GIM aims at responding to climate change through a combination of adaptation and mitigation measures.

A. Mission Objective: To use a combination of adaptation and mitigation measures in enhancing carbon sinks in sustainably managed forests and other ecosystems, adaptation of vulnerable species/ecosystems, and adaptation of forest-dependent communities.

B. Mission Targets and Timeline: To achieve its objective, the mission targets are as under:

- Increase forest/tree cover on 5 million hectares of forest/non-forest lands and improve quality of forest cover on another 5 million hectares.
- Improve ecosystem services including biodiversity, hydrological services and carbon sequestration through

treatment of an area of 10 million hectares.

- Increase forest-based livelihood income of about 3 million households living in and around the forests.
- Enhance annual CO_2 sequestration by 50 to 60 million tonnes in the year 2020.

The implementation of the mission will spread over 10 years.

4.9.7 National Mission on Sustainable Habitat (NMSH): NMSH seeks to promote sustainability of habitats through the following measures:

A. Mission Objective: To promote sustainability of habitats through improvements in energy efficiency in buildings, urban planning, improved management of solid and liquid waste including recycling and power generation, modal shift towards public transport and conservation.

B. Mission Targets and Timeline: To achieve its objective, the mission targets are as under:

- Extension of the Energy Conservation Building Code, which addresses the design of new and large commercial buildings to optimise their energy demand. Incentives will be provided for re-tooling existing building stock.
- Better urban planning and modal shift to public transport by making long-term transport plans to facilitate the growth of medium and small cities in such a way that ensures efficient and convenient public transport.
- Recycling of material and urban waste management under which a special area of focus will be development of technology for producing power from waste. The mission will include a major R&D programme, focussing on bio-chemical conversion, waste water use, sewage utilisation and recycling options wherever possible.

The mission includes timelines for all the strategies and sub-components of each strategy. These range between 2009 and 2017.

4.9.8 National Mission for Sustainable Agriculture (NMSA): NMSA seeks to address issues regarding sustainable agriculture in the context of risks associated with climate change by devising appropriate adaptation and mitigation strategies for ensuring food security, enhancing livelihood opportunities, and contributing to economic stability at national level. Under NMSA, the adaptation and mitigation measures would be mainstreamed in the following areas:

A. Mission Objective: To transform agriculture into an ecologically sustainable climate resilient production system while, at the same time, exploiting its fullest potential and thereby ensuring food security, equitable access to food resources, enhancing livelihood opportunities and contributing to economic stability at the national level.

B. Mission Targets and Timeline: To achieve its objective, the mission will work on the following major programme components or activities:

- **Rain-fed Area Development:** Adopt an area-based approach for development and conservation of natural resources along with farming systems.
- **On-Farm Water Management:** Enhance water use efficiency by promoting efficient on-farm water management technologies and equipment.
- **Soil Health Management:** Promote location as well as crop-specific sustainable soil health management.
- **Climate Change and Sustainable Agriculture Monitoring, Modelling and Networking:** Creation and bi-directional (farmers to research institutions and vice versa) dissemination of climate change related information and knowledge.

While promotion of dry land agriculture would receive prime importance by way of developing suitable drought- and pest-resistant crop varieties and ensuring adequacy of institutional support, NMSA would also expand its coverage to rain-fed areas for integrating farming systems with livestock and fisheries so that agriculture continues to grow in a sustainable manner.

In short, NAPCC identifies a number of measures that simultaneously advance the country's development and climate change related objectives of adaptation and mitigation. The implementation of the NAPCC is designed to take place through eight National Missions, which form the core of the National Action Plan and incorporate multi-pronged, long-term and integrated strategies for achieving India's key goals in the context of climate change.

4.10 Other National and Sub-National Initiatives

In addition to the National Action Plan on Climate Change, the Government of India has taken several other measures to promote sustainable development and address the threat of climate change. These initiatives operate at the national and sub-national level and span domains that include climate change research, clean technology research and development, finance, and energy efficiency and renewable energy policy and deployment.

4.10.1 National Clean Energy Fund (NCEF): The Government of India created NCEF in 2010 for the purpose of financing and promoting clean energy initiatives and funding research in the area of clean energy in the country. The corpus of the fund is built by levying a cess of ₹ 50 (subsequently increased to ₹ 100 in 2014) per tonne of coal produced domestically or imported.

4.10.2 State Action Plan on Climate Change: In 2008, the Government of India launched the National Action Plan on Climate Change (NAPCC) establishing eight priority missions. For the realisation of these proposed actions at the sub-national level, in August 2009 the Prime Minister called upon State Governments to prepare their own State Action Plan on Climate Change (SAPCC) consistent with strategies in the NAPCC.

A common framework for the preparation of SAPCC was developed to harmonise national and state level actions. The common framework drew largely on the principles of territorial approach to climate change which focused on sub-national planning, building capacities for vulnerability assessment and identifying investment opportunities based on the priorities of a state. The framework provided broad, systematic and step-wise process for the preparation of SAPCC and advocated a participatory approach so that states have enough ownership for the process and the final plan. The recommended approach retained a level of flexibility in order to integrate state level variations in ecosystems, geographic conditions, socio-economic scenario, and other factors.

Almost all states have prepared their State Action Plan. The SAPCCs have both adaptation and mitigation component to address climate change impacts, though adaptation has been identified as a more important element of the plan. These plans have been reviewed and endorsed by the Expert Committee on Climate Change under the Ministry of Environment, Forest and Climate Change (MoEFCC), Government of India.

4.10.3 NABARD's Role in Combating Climate Change: India's National Bank for Agriculture and Rural Development (NABARD) has been accredited as a National Implementing Entity (NIE) for India for the Adaptation Fund created under the United Nations Framework Convention on Climate Change (UNFCCC). At present, NABARD is the only NIE in the Asia-Pacific Region.

In its capacity as NIE, NABARD has generated several feasible projects on climate change adaptation in diverse agro-climatic regions and livelihood sectors, five of which have been submitted as proposals to the Adaptation Fund amounting to US$ 7.3 million. The Adaptation Fund Board (AFB) in its 24th meeting held on October 9, 2014, sanctioned the first set of two projects submitted by NABARD with an outlay of US$ 3.2 million for promoting climate resilient agriculture systems in West Bengal and enabling the fisheries sector in Andhra Pradesh to respond to the challenge posed by sea level rise.

Additionally, NABARD is implementing several developmental projects to promote sustainable livelihoods through natural resource management (NRM) such as watershed development and sustainable livelihood for tribal communities, which are helping build climate change resilience and adaptive capacities of rural communities. NABARD has sanctioned a pilot project of ₹ 21 crore on climate change adaptation in Ahmednagar (Maharashtra) to develop knowledge, strategies and approaches that will enable vulnerable communities to adapt to the impending impacts of climate change such as delays in onset and withdrawal of the monsoon, changes in temperature and unseasonal rainfall. Pilot projects on climate proofing of watersheds in Tamil Nadu and Rajasthan are also underway.

NABARD has also financed projects that improve energy efficiency. In the state of Karnataka, energy inefficient pumps were replaced by more efficient ones through a loan assistance programme. Under its Infrastructure Development Assistance (NIDA), NABARD is financing green investments on solar power generation and improvement of electricity distribution networks, which include India's first one MW canal-top solar power project in the state of Gujarat.

NABARD is collaborating with non-governmental organisations, research institutions and government departments to conduct training programmes and workshops. These are building the capacities of NABARD's partners and stakeholders to understand the adverse impacts of climate change on natural and human systems and to design appropriate adaptation measures.

The experience gained by NABARD as NIE has enabled the institution to adopt processes and procedures that are followed internationally in the development, review and sanctioning of projects. NABARD continues to develop quality projects to create a pipeline of feasible projects that will help communities in India take measures to adapt to the impacts of climate change.

4.10.4 Prime Minister's Council on Climate Change (PMCCC): Government of India set up an elaborate institutional mechanism to consider and address issues relating to climate change. A Council chaired by Prime Minister called Prime Minister's Council on Climate Change was constituted in June 2007 to coordinate national action for assessment, adaptation and mitigation of climate change. The Council provides the overall guidance to climate change related actions taken by various Ministries of the Government and other agencies. The Council was reconstituted in November 2014 with the following objectives:

1. Co-ordinate national action plans for assessment, adaptation and mitigation of climate change.
2. Advise government on pro-active measures that can be taken by India to deal with the challenge of climate change.
3. Facilitate inter-ministerial coordination and guide policy in relevant areas.

PMCCC has directed the missions under the NAPCC to enhance their ambition in respect of adaptation, mitigation and capacity building and reprioritize them, besides recommending the setting up of some new missions in addition to the existing eight. Considering the adverse impacts that climate change could have on health, a new Mission on Climate Change and Health is currently under formulation and a National Expert Group on Climate Change and Health has been constituted. The proposed waste-to-energy mission will incentivize efforts towards harnessing energy from waste and is aimed at lowering India's dependence on coal, oil and gas for power production. The National Mission on Coastal Areas (NMCA) will prepare an integrated coastal resource management plan and map vulnerabilities along the entire (nearly 7,000 km. long) shoreline.

The Ministry of Earth Sciences will provide it scientific and technical advice and the Ministry of Environment, Forest and Climate Change (MoEFCC) will manage and implement the NMCA.

4.10.5 Indian Network for Climate Change Assessment (INCCA): To enhance knowledge about the impacts of climate change at the national and sub-national level, INCCA was launched on October 14, 2009.

INCCA has been conceptualised as a network-based scientific programme designed to:

- Assess the drivers and implications of climate change through scientific research.
- Prepare climate change assessments once every two years (greenhouse gas estimations and impact of climate change, associated vulnerabilities and adaptation).
- Develop decision support systems.
- Build capacity towards management of climate change related risks and opportunities.

The programmes envisaged under the aegis of INCCA include those on Black Carbon, Ecosystem Monitoring, Centre for Advanced Studies, Impact Assessments, Greenhouse Gas Inventory Programme, Integrated Vulnerability and Adaptation (V&A) Assessments and Developing Scenarios.

Under the Greenhouse Gas Inventory Programme, a report on 'India's Greenhouse Gas Emission, 2007' was published in May 2010. This report contains updated information on India's greenhouse gas emissions for the year 2007. The assessment was prepared by 83 experts belonging to 19 institutions from across the country.

Under the Impact Assessments Programme, a "Climate Change in India: 4x4 Assessment" has been undertaken to ascertain impacts in 2030s. This assessment brings together four major regions in India, namely Himalayan region, the North-Eastern region, the Western Ghats and the Coastal Region in regard to observed climate and climate change projections in the year 2030 on 4 key sectors of agriculture, water, natural ecosystem, biodiversity, and health.

The National Carbonaceous Aerosols Programme (NCAP) aims to enhance the understanding of the role of carbonaceous aerosols on climate change, to prepare an inventory of black carbon emissions in the country and to assess its impacts on glacier melting. The science plan for the "Black Carbon Research Initiative—National Carbonaceous Aerosols Programme (NCAP)" was launched in March 2011.

4.10.6 National Institute for Climate Change Studies and Actions (NICCSA): NICCSA is being set up by the Government of India under the Climate Change Action Programme (CCAP) of the Ministry of Environment, Forest and Climate Change (MoEFCC). The Institute will conduct analytical studies on scientific, environmental, economic development and technological issues related to climate change.

4.10.7 Expert Group on Low Carbon Strategies for Inclusive Growth: This Expert Group was set up by the Planning Commission, Government of India, in 2010 to suggest low carbon pathways consistent with inclusive growth in India. The high-level Expert Group was composed of representatives from relevant government ministries, industry, think tanks and research institutions.

The Expert Group was required to report on the following:

- Its evaluation of some key alternative low carbon options with an analysis of their cost-benefit, and relative merits and demerits.
- An Action Plan comprising of critical low carbon initiatives to be undertaken, including sector-specific initiatives, along with a suggested timeline and targets starting 2011, that can feed into the Twelfth Five Year Plan (2012-17).
- List of enabling legislations, rules or policies required to operationalise the low carbon roadmap.

The Expert Group submitted its Interim Report in May 2011. This report provided a menu of options and showed that it is possible for India to reduce its emission intensity by 20-25 percent over 2005 levels by the year 2020. The Final Report, submitted in April 2014, provides a more detailed and longer term assessment of these options, and the macro-economic and welfare implications of the low carbon strategy.

The Expert Group gave high priority to inclusive growth and noted that pursuit of low carbon development is consistent with growth and inclusion. In the low carbon strategy assessed, energy efficiencies in households, buildings, industry and transport play important roles. At the same time, low carbon supply technologies, such as solar and wind in the power sector, and greater use of public transport and non-motorized transport are critical. Increased sequestration through enlarged green cover through Green India Mission also helps. Reduced local environmental pollution and dependence on imported energy were also identified as benefits of a low carbon strategy. However, the Expert Group recognised that the additional investment required for a low carbon strategy would leave less investment available for other sectors.

To sum up, India is extremely vulnerable to the impacts of climate change and significant measures are needed to build climate resilience and assist communities with adaptation. Several programmes are already underway in many parts of India, often in partnership with local financial institutions and grassroots non-governmental organisations that are working with local communities on project implementation. Through its corporate social responsibility programmes and other initiatives, Indian industry is also promoting sustainable livelihoods and infrastructure development across the country. Additionally, there are measures that India can take to promote sustainable and inclusive growth in a less carbon-intensive manner. Many of these actions will have to be designed and implemented by industry and civil society organisations.

The change in the climate system is likely to have adverse impacts on livelihoods, cropping pattern and food security. Extreme heat events are likely to be longer and more intense in addition to changes in the precipitation patterns. Adverse impacts are likely to be felt more acutely in tropical zone countries such as India, and within India, the poor will be more exposed.

India can make a significant contribution in addressing climate change. Unlike some countries, it has taken substantial actions to eliminate petroleum subsidies and gone beyond to impose substantial taxes on these products. These actions have taken India from a carbon subsidization regime to one of significant carbon taxation regime—from a negative price to an implicit positive price on carbon emissions.

Climate change needs to be accounted as one of the potential risk factors affecting the outcome of any development programme or process in the long-run. Specific institutional arrangements are needed within government to oversee and monitor development in the context of the above necessities.

Forecasting systems and adaptations in response to climate change should be the key concerns for future development. For living systems, study of the impact of factors considered subject to rapid changes and their adaptation potential against the changing factors will be important. For predictions and early warning systems, understanding of the processes which could show discernible changes well before the actual impact starts is essential.

Vulnerabilities of India to climate change—in particular impacts on water resources, forests, coastal areas, agriculture, and health—should be identified. For this, co-ordinated efforts through the relevant scientific institutions, in both government and non-government sectors, are required. For specific thematic areas, nodal institutions can be identified and designated.

Initiatives are needed in the area of adaptation to climate change. The extension of action needed obviously depends upon the expected success in mitigation of climate change and this is difficult to predict at this stage. However, since a substantial adverse change appears unavoidable at present, there is need to embark on a process of adaptation which would, at least, identify the initiatives needed to deal with those outcomes that are highly likely. This requires action in the area of agricultural research to evolve varieties that can cope with likely climate changes.

Reconciling India's climate change goals and energy imperatives will require a major technological breakthrough to make the burning of coal cleaner and greener. Areas for action by India include initiatives in clean energy including renewable energy and action to increase energy efficiency.

India, as a party to UNFCCC, 1992 has been committed to address the global problem on the basis of the principle of *common but differentiated responsibilities and respective capabilities* of the member Parties. Since India is highly vulnerable to climate change, it is in its interest to have a global agreement. This, however, has to be consistent with its National Environmental Policy, 2006 which requires *equal per capita entitlements of global environmental resources to all countries*.

5

Land Degradation and Droughts

5.1 Land Pollution and Degradation

Land, a non-renewable resource, is central to all primary production systems. Over the years, India's landmass has suffered from different types of degradations. Floods and droughts are recurrent phenomena in India, severely constraining agriculture. The problem of land degradation gets exacerbated due to large scale land erosion in which fertile top soil is lost.

As per study conducted by National Bureau of Soil Survey and Land Use Planning, Nagpur, [a Regional Centre of Indian Council of Agricultural Research (ICAR)], an area of about 146.82 million hectares is subjected to various kinds of land degradation, viz. water and wind erosion, water logging, alkalinity/salinity, soil acidity and other complex problems across the country. Frequent events of droughts, floods and climate change also have an impact on soil fertility and land degradation, thereby affecting foodgrains production in the country.

5.1.1 Causes: Land degradation occurring due to the natural and human induced causes, like wind erosion and water logging, is one of the priority concerns in India. The varying degrees and types of degradation stem mainly from unsustainable use and inappropriate land management practices.

A. Intensive Agricultural Practices: Intensive agricultural practices that rely heavily on water, chemical fertilisers and pesticides have caused water-logging and salinity in many parts of the country. The expansion of the irrigation system without adequate steps for treatment of the catchment areas has exacerbated this. The quest for increased agricultural productivity has led to intensive cultivation of marginal lands causing their degradation.

B. Unsanitary Disposal of Solid Wastes: Pollution of

land results also from the unsanitary disposal of solid wastes. In India open dumping of such wastes of municipal and industrial origin on low lying land is a common phenomenon. This serves as a breeding ground for pests and disease carrying vectors.

C. Pressure of Human and Cattle Population: In addition to the wastelands, other areas such as deserts, drought-prone, flood-prone and tribal areas have been subjected to severe forms of degradation. The capacity of these lands is limited due to environmental factors. Pressures of human and livestock population have further compromised them.

D. Community Lands: Some of the most degraded lands in the country are the common property resources (CPRs). CPRs are resources on which people have an equal right of use. These resources include community pastures, community forests, wastelands and common dumping and threshing grounds. In spite of concerted efforts to check deforestation and the various afforestation schemes taken up during successive Five Year Plans, large tracks of forest continue to be classified as degraded.

E. Mining Activities: Land degradation is considered to be unavoidable by-product of mining and has reached alarming proportions, mainly due to over-exploitation and mismanagement of natural resources. Mining activity often leads to environmental problems like land degradation, particularly in opencast mining and land subsidence in underground mining. Open-cast mining in areas with forest cover causes deforestation.

F. Industrial Effluents: Soil pollution from heavy metals due to improper disposal of industrial effluents, along with the excessive use of pesticides and mismanagement of domestic and municipal wastes, is becoming a major concern. Though no reliable estimates are available to depict the exact extent and degree of this type of land degradation, it is believed that the problem is extensive and its effects are significant. Some commercial fertilizers also contain appreciable quantities of heavy metals, which have undesirable effects on the environment. The indiscriminate use of agrochemicals, such as fertilizers and pesticides, is often responsible for land degradation.

G. Over Extraction from Forests: Loss of vegetation occurs as a result of deforestation, cutting beyond the silviculturally permissible limits, unsustainable fuel-wood and fodder extraction, shifting cultivation, encroachment into forest lands, forest fires and overgrazing, all of which subject the land to degradational forces.

H. Other Causes: Other important factors responsible for large-scale degradation are the following:

1. Non-adoption of adequate soil conservation measures.
2. Improper crop rotation.
3. Indiscriminate use of agro-chemicals such as fertilizers and pesticides.
4. Improper planning and management of irrigation systems.
5. Extraction of groundwater in excess of the recharge capacity.
6. Extension of cultivation to lands of low potential or high natural hazards.

In addition, there are a few underlying or indirect pressures such as land shortage, short-term or insecure land tenancy, open access resource, economic status and poverty of the agriculture dependent people which are also instrumental, to a significant extent, for the degradation of land.

5.2 Consequences of Land Degradation

Land degradation has a direct bearing on the productivity of soil, its vulnerability to rainfall variations, scarcity of drinking water, fodder and fuel wood. Given the inter-linkages of crop production, livestock economy and environment, land degradation has a major impact on the livelihoods of the people, especially in rural areas.

The increasing generation of solid waste is a growing problem in all cities. Uncollected garbage is not only visually ugly, it also causes diseases. Dumping garbage into landfills is not an attractive option in land scarce cities and should be minimised. Most industrial countries now require their citizens to segregate wastes at home into recyclable products. Cities and towns must be encouraged to evolve systems with citizen participation, for segregation of waste at point of origin, maximising recycling and safe disposal of the rest. India's villages also have a huge problem of garbage disposal which must be addressed through self-help groups (SHGs) and panchayti raj institutions (PRIs).

Direct consequences of agricultural development on the environment arise from intensive farming activities, which contribute to soil erosion, land salination and loss of nutrients. The introduction of *green revolution* in the country has been accompanied by over-exploitation of land and water resources and excessive usage of fertilizers and pesticides. Shifting cultivation has also been a major factor responsible for land degradation in hilly areas. Leaching due to extensive use of pesticides and fertilizers is a major source of contamination of water bodies.

The extent of agricultural intensification and extensification is characterized by an increase in cropping and irrigation intensity and the unbalanced use of chemical fertilizers, pesticides and insecticides. It has also led to land degradation, overexploitation of underground water resources and increased use of chemical fertilizers, leading to eutrophication and water pollution in some regions.

Enhanced intensification and extensification also leads to salination, alkalization and water logging in irrigated areas, along with eutrophication of water bodies and ill health of oceans, leading to loss of biodiversity. For achieving and maintaining food security and sustainable forestry, controlling of land/soil erosion is extremely vital. It is essential to control soil erosion in order to attain and maintain food security, sustainable forestry and agricultural and rural development.

5.3 Strategies to Check Land Degradation

The prevention of land degradation and the augmentation of the carrying capacity of land to provide food, fuel and fodder requirements have, therefore, been a primary concern of the Government. Special Area Development Programmes have been funded to mitigate the harsh living conditions of people through better land management, water harvesting and conservation practices on a watershed basis.

India has paid a good deal of attention to harness its water resources by way of construction of major, medium and minor irrigation projects and the development of ground water resources. Organizations have also been set up in this field, namely Central and State Irrigation Departments, Central Water Commission and the Central Ground Water Board. However, very little attention has been paid to the proper management of land and soil resources with the result that they have suffered very serious degradation. The losses which the country is bearing on account of the continuing degradation of its land resources are of staggering dimensions and constitute one of the important threats to India's economic progress.

Land degradation problem could be tackled to an extent by suitable policies that would internalize the issue into proper decision-making. At the macro level, the existing database on land use statistics cannot adequately facilitate the analysis of land degradation and its impact. Changes in the classification of land use statistics are needed in order to study its impact. Advanced technology like remote sensing could go a long way in generating vital information on different dimensions of land degradation.

The information base on which farmers make decisions is incomplete in terms of internalizing rapid changes in soil and water quality variables. Hence, there is the need to move towards more sustainable practices such as integrated pest management and land-conserving crop rotations. Research needs to be focused on measures such as integrated crop management. An integrated approach to the problem of degradation, linking agriculture and environment, is yet to be attempted even at the policy level.

Overuse of resources has been identified as a major cause of land degradation. Creation of forest/biomass resources in all the culturable vacant lands should be taken up for strengthening life support system of communities and maintaining soil/water regimes.

5.4 Droughts: Causes, Consequences and Remedial Strategies

Major famines in recorded history have taken a heavy toll of many millions. Each of them caused tremendous suffering and depopulated particular regions due to starvation deaths.

During the colonial period, many droughts turned into severe famines causing massive human losses. According to one estimate, in the latter half of the 19th century, there were approximately 25 major famines across India, which killed 30-40 million people. The first Bengal famine of 1770 is estimated to have wiped out nearly one-third of the population. The famines continued until Independence in 1947, with the Bengal famine of 1943-44 being among the most devastating, affecting 3-4 million.

The situation improved remarkably in post-Independent India. The Green Revolution in the 1960s made the country self-sufficient in food production. Though India's population has tripled since it achieved independence, there has been no famine in the past 50 years, which is certainly an impressive achievement.

In the 1990s, with the liberalization of the Indian economy and consequent accelerated growth in industry and services, the share of agriculture in gross domestic product (GDP) has fallen to barely 13 percent. As a result, the country has become far more resilient in absorbing the impact of drought at the macro-economic level. Though the drought of 2002 was the severest, its impact on the economy was marginal. Key indicators, such as inflation and wholesale price index, were not adversely affected and conversely, the foreign exchange reserves reached an unprecedented level of US$ 64 billion.

In recent times there have been no famines in India mainly because of the availability of a buffer stock of foodgrains, better transportation and storage facilities as well as greatly improved logistics because of application of technology. Over and above, the Government of India has adopted food security as a prime objective of public policy.

5.4.1 Drought Defined: It is difficult to provide a precise and universally accepted definition of drought due to its varying characteristics and impacts across different regions of the world, such as rainfall patterns, human response and resilience, and diverse academic perspectives.

Drought is a temporary aberration unlike aridity, which is a permanent feature of climate. Seasonal aridity (i.e. a well-defined dry season) also needs to be distinguished from drought. Thus drought is a normal, recurrent feature of climate and occurs in all climatic regimes and is usually characterized in terms of its spatial extension, intensity and duration. Conditions of drought appear when the rainfall is deficient in relation to the statistical multi-year average for a region, over an extended period of a season or year, or even more.

Drought differs from other natural hazards—such as cyclones, floods, earthquakes, and volcanic eruptions—in following respects:

1. Being of slow-onset it is difficult to determine the beginning and end of the event.
2. Duration may range from months to years and the core area or epicentre changes over time, reinforcing the need for continuous monitoring of climate and water supply indicators.
3. No single indicator or index can identify precisely the onset and severity of the event and its potential impacts; multiple indicators are more effective.
4. Spatial extent is usually much greater than that for other natural hazards, making assessment and response actions difficult.

5. Impacts are cumulative and the effects magnify when events continue from one season or year to the next.

5.4.2 Causes of Recurrence of Droughts in India: A deficiency in rainfall causes depletion of both surface and ground water levels and adversely affects agricultural operations. The conditions for onset of drought in India vary across agro-climatic zones. In the semi-arid regions, even a 400 mm rainfall would be adequate for the growth of crops, while in high rainfall regions of Assam, even an annual rainfall of 1,000 mm would create conditions for drought.

Drought is a recurrent climatic phenomenon in India and is caused due to the country's peculiar physical and climatic characteristics as well as resulting economic and agricultural impacts. These include the following:

1. High average annual rainfall of around 1,150 mm; no other country has such a high annual average; however, there is considerable annual variation.
2. About 73 percent of the total annual rainfall is received in less than 100 days during the south-west monsoon and the geographic spread is uneven.
3. Even though India receives abundant rain as a whole, disparity in its distribution over different parts of the country is so great that some parts suffer from perennial dryness. In other parts, however, the rainfall is so excessive that only a small fraction can be utilized. The variability in rainfall as compared to long period average (LPA) exceeds 30 percent in large areas of the country and is over 40-50 percent in parts of drought-prone Saurashtra, Kutch, and Rajasthan.
4. Around 33 percent of the cropped area in the country receives less than 750 mm rain annually making such areas hotspots of drought.
5. Inadequacy of rains coupled with adverse land-man ratio compels the farmers to practice rain-fed agriculture in large parts of the country.
6. Irrigation, using groundwater aggravates the situation in

the long term as groundwater withdrawal exceeds replenishment; in the peninsular region availability of surface water itself becomes scarce in the years of rainfall insufficiency.

7. Per capita water availability in the country is steadily declining.
8. Traditional water harvesting systems have been largely abandoned.

5.4.3 Classification of Droughts: In the literature, droughts have been classified into three categories in terms of impact:

1. Meteorological drought.
2. Hydrological drought.
3. Agricultural drought.

A. Meteorological Drought: It is defined as the deficiency of precipitation from expected or normal levels over an extended period of time. Meteorological drought usually precedes other kinds of drought. According to the legend, meteorological drought is said to occur when the seasonal rainfall received over an area is less than 25 percent of its long-term average value. It is further classified as moderate drought if the rainfall deficit is 26-50 percent and severe drought when the deficit exceeds 50 percent of the normal.

B. Hydrological Drought: It is best defined as deficiencies in surface and subsurface water supplies leading to a lack of water for normal and specific needs. Such conditions arise, even in times of average (or above average) precipitation when increased usage of water diminishes the reserves.

C. Agricultural Drought: Triggered by meteorological and hydrological droughts, it occurs when soil moisture and rainfall are inadequate during the crop growing season causing extreme crop stress and wilting. Plant water demand depends on prevailing weather conditions, biological characteristics of the specific plant, its stage of growth and the physical and biological properties of the soil.

Thus, agricultural drought arises from variable susceptibility of crops during different stages of crop development, from emergence to maturity. In India, it is defined as a period of four consecutive weeks (of severe meteorological drought) with a rainfall deficiency of more than 50 percent of the long-term average (LTA) or with a weekly rainfall of 5 cm or less from mid-May to mid-October (the kharif season) when 80 percent of India's total crop is planted or six such consecutive weeks during the rest of the year.

5.4.4 India-specific Classification of Droughts: A drought year as a whole is defined as a year in which, the overall rainfall deficiency is more than 10 percent of the long period average value (LPA) value and more than 20 percent of its area is affected by drought conditions, either moderate or severe or combined moderate and severe. The country experienced all-India drought years in 1972, 1979, 1987 and 2002.

Droughts are also classified according to the timing of rainfall deficiency during a particular rainfall season, usually June to September in India. An analysis of droughts in recent decades indicates that in 1972, 1987 and 2002, rainfall deficiency was the maximum during July. These are called early season droughts and provide sufficient lead-time to mitigate the impact of drought. In 1965 and 1979, rainfall deficiency in September was the major cause of drought. These droughts are called late season droughts. Mid-season droughts occur in association with the breaks in the southwest monsoon. If the drought conditions occur during the vegetative phase of crop growth, it might result in stunted growth, low leaf area development and even reduced plant population.

In India, around 68 percent of the country is prone to drought in varying degrees. Of the entire area, 35 percent of the area, which receives rainfall between 750 mm and 1,125 mm, is considered drought-prone, while another 33 percent, which receives less than 750 mm of rainfall, is called chronically drought-prone. A further classification of India's regions can be done into arid (19.6 percent), semi-arid (37 percent), and sub-humid areas (21 percent).

5.4.5 Nature of Droughts in India: In India, the occurrence and conditions of drought are influenced by a number of factors. Rainfall and cropping patterns are different across many geographical regions. It is not just the deficiency of rainfall, but also the uneven distribution of rainfall across the season, duration of rainfall deficiency and its impact on different regions of the country that characterize drought conditions. The facts presented below show seasonal and regional characteristics of drought conditions in India. The information on these aspects of droughts is necessary for devising and implementing an effective mitigation programme.

A. Seasonal Characteristics and Intra-Seasonal Variability: India receives most of its rainfall (73 percent) from the south-west or "summer" monsoon (the rainfall received between June and September). The performance of the Indian economy is vitally linked with the rainfall that occurs during these months. The summer monsoon sets in during the first week of June in the south-east corner of India and gradually proceeds towards the north-west region covering the entire country by the second week of July. Monsoon starts its withdrawal during the first week of September from the west and north and gradually recedes from the entire country.

Due to this pattern of onset and withdrawal, the north-west region receives less than a month of rainy season due to late arrival and early cessation of monsoon conditions. Conversely, Kerala and North-Eastern parts of India receive more than 4 months of rainfall due to the early arrival and later withdrawal of the monsoon.

Coastal areas of peninsular India and Tamil Nadu, in particular, also receive rains from October to December, primarily due to periodic cyclonic disturbances in the Bay of Bengal (north-east monsoon or post-monsoon). It also receives some rainfall during other months.

B. Duration of the Drought: Drought conditions are prolonged due to poor rainfall in consecutive years. Prolonged drought conditions do not allow people the opportunity to recover from the impact of drought and it depletes their capacity to cope with the impact.

C. Distribution of Drought-prone Areas: In India, drought-prone areas comprise a total land area of 329 million hectares, with three-fourths being arid, semi-arid and sub-humid areas.

(a) Arid zone (19.6 percent): Mean annual precipitation (MAP) of 100-400 mm (water deficit throughout the year); Rajasthan, parts of Haryana and Gujarat. Droughts are severe in this zone.

(b) Semi-arid zone (37.0 percent): MAP of 400-600 mm (water surplus in some months and deficit in other months); parts of Haryana, Punjab, west Uttar Pradesh, west Madhya Pradesh, and also most of the peninsular parts of the Western Ghats. Drought can be moderate to severe in this zone.

(c) Dry sub-humid zone (21.0 percent): MAP of 600-900 mm in India; parts of northern plains, central highlands, eastern plateau, parts of Eastern Ghats and plains and parts of western Himalayas. Droughts are moderate in this zone.

(d) Humid and Per-humid Regions: Such as Assam and other North-Eastern States which rarely face drought.

A drought situation arising in any part of India needs to be assessed in terms of the overall seasonal rainfall and intra-seasonal distribution of rainfall. Further, the period over which the deficiency of rainfall has taken place would be a critical indicator of the severity of drought. Consecutive drought years would aggravate the impact of drought.

Finally, the drought management measures would largely depend upon the area in which the drought has occurred. Relief as well as mitigation measures would be different for chronically drought prone areas, compared to dry sub-humid areas. The drought monitoring system also needs to be adapted to these differentiated features of drought.

5.4.6 Impacts of Droughts: Drought produces wide-ranging impacts that span many sectors of the national economy. These impacts are felt much beyond the area experiencing physical drought. The complexity of these impacts arises because water is integral to our ability to produce goods and provide services. Drought produces both direct and indirect impacts.

Direct impacts or primary impacts are usually physical/material and include reduced agricultural production; increased fire hazard; depleted water levels; higher livestock and wildlife mortality rates; and damage to wildlife and fish habitats.

When direct impacts have multiplier effects through the economy and society, they are referred to as indirect impacts. These include a reduction in agricultural production that may result in reduced income for farmers and agribusiness, increased prices for food and timber, unemployment, reduced purchasing capacity and demand for consumption, default on agricultural loans, rural unrest, and reduction in agricultural employment leading to migration and drought relief programmes. The more removed the impact from the cause, the more complex is the link to the cause. These multiplier effects are often so diffuse that it is very difficult to generate financial estimates of actual losses caused by a drought.

The impacts of drought are generally categorized as economic, environmental, and social.

A. Economic Impacts: These refer to production losses in agriculture and related sectors, especially forestry and fisheries, because these sectors rely on surface and subsurface water supplies. It causes a loss of income and purchasing power, particularly among farmers and rural population dependent on agriculture. All industries dependent upon the primary sector for their raw materials would suffer losses due to reduced supply or increased prices. Drought thus has a multiplier effect throughout the economy, which has a dampening impact on employment, flow of credit and tax collections. If the drought is countrywide, macroeconomic indicators at the national level are adversely impacted.

B. Environmental Impacts: These include lower water

levels in reservoirs, lakes and ponds as well as reduced flows from springs and streams would reduce the availability of feed and drinking water and adversely affect fish and wildlife habitat. It may also cause loss of forest cover, migration of wildlife and their greater mortality due to increased contact with agricultural producers as animals seek food from farms and producers are less tolerant of the intrusion. A prolonged drought may also result in increased stress among endangered species and cause loss of biodiversity.

Reduced stream flow and loss of wetlands may cause changes in the levels of salinity. Increased groundwater depletion, land subsidence, and reduced recharge may damage aquifers and adversely affect the quality of water (e.g., salt concentration, increased water temperature, acidity, dissolved oxygen, turbidity). The degradation of landscape quality, including increased soil erosion, may lead to a more permanent loss of biological productivity of the landscape.

C. Social Impacts: These arise from lack of income causing out migration of the population from the drought-affected areas. People in India seek to cope with drought in several ways which affect their sense of well-being: they withdraw their children from schools, postpone daughters' marriages, and sell their assets such as land or cattle. In addition to economic hardships, it causes a loss of social status and dignity, which people find hard to accept. Inadequate food intake may lead to malnutrition, and in some extreme cases, cause starvation. Access and use of scarce water resources generate situations of conflict, which could be socially very disruptive. Inequities in the distribution of drought impacts and relief may exacerbate these social tensions further.

5.4.7 Strategies for Drought Mitigation: A drought makes a very perceptible impact on populations that are largely dependent upon agriculture and related occupations for their livelihood. As crops are adversely affected, agricultural income shrinks and causes loss of employment in the agriculture sector. It also has an indirect impact on the other sectors of economy. On the supply side, drought causes a shortage of raw material supplies for agro-based industries and on the demand side, it reduces the demand for industrial products due to diminished purchasing capacity of the rural consumers. Most major droughts in India were followed by recession. Though drought makes its impact over time, it poses a serious challenge for human well being.

Strategies for drought mitigation comprise the following elements:

1. Drought risk and vulnerability assessment.
2. Forecasting and early warning.
3. Structural mitigation measures.
4. Non-structural mitigation measures.
5. Awareness generation.
6. Research and development.
7. Community participation.

A. Vulnerability and Risk Assessment: State Governments are primarily responsible for mitigation of drought. State Governments should set up a mission/task force on drought mitigation for advising them on policies and programmes. The said mission/task force should also conduct drought risk and vulnerability assessment for identifying drought prone areas, nature and severity of drought, vulnerable economic sectors, communities and individuals etc. Such an assessment will provide an economic rationale for interventions and in identifying mitigation measures. A composite risk and vulnerability assessment would help in developing long-term policies and programmes for drought risk mitigation at the State level and in framing appropriate strategies for addressing the critical gaps during drought relief. As a part of this exercise, States would need to prepare vulnerability maps especially in the case of rain-fed areas in order to bring them under assured irrigation to the extent possible.

B. Forecasting and Early Warning: The Crop Weather Watch Group (CWWG), an Inter-Ministerial forum at Central level, meets on regular basis to take stock of rainfall, weather forecast, progress of sowing, crop health, storage level in major water reservoirs, pest control, inputs availability etc. Observations of CWWG are shared among Central Ministries/Departments and State Governments for formulating appropriate strategy to meet drought related contingencies, if any. Efforts of State Governments are supplemented from Central resources, whenever the situation warrants for mitigating hardships in agriculture and allied sectors.

C. Structural Mitigation Measures: Structural mitigation measures include involvement of Central and State Governments by providing technological solutions to drought mitigation primarily through developmental programmes and relief operations.

Action Plan for revisiting on-going and formulating future programmes/schemes design of any future developmental programmes will invariably have to be compatible with the imperatives of drought resilience. Such an approach will naturally vary for different agro-climatic zones depending upon their vulnerability profile and thus be selective and priority driven.

It will also be imminent to pursue convergence of various developmental schemes and programmes at both Central and State levels to minimise duplicity of efforts and investment in order to derive maximum results in managing drought with available resources. It is imperative to establish a dedicated drought monitoring cell (DMC) at State level with requisite communication network to facilitate dissemination of the required information through agricultural extension agencies, kisan call centres (KCC), media advisories, web portal etc. The control room/cells at Central and State Government levels may remain functional throughout the year and they can be suitably strengthened for effective management of drought. Drought being directly associated with climatic variability, it will be useful to strengthen existing institutional framework for monitoring, assessment and dissemination of real time forecast on drought.

D. Non-Structural Mitigation Measures: The non-structural measures envisage enhancing management skills to improve preparedness of vulnerable communities, panchayati raj institutions (PRIs), urban local bodies and State authorities to mitigate the risks and respond effectively to drought situations. The following non-structural measures may be considered to facilitate this:

- Hazard-zoning by the State Governments, i.e. identification of drought prone areas by using *vulnerability mapping.*
- Preparation of inventory of resources by the State governments to meet likely emergency requirements of inputs (seeds, fertilizers etc.) and energy (power and diesel for pump sets), tankers and wagons (for drinking water movement), fodder availability (planning for movement from surplus areas including those from forest areas), availability of food grains and advance allocation plan, agricultural credit, planning for risk insurance etc.
- Knowledge networking of best practices both at the State and Central levels (e.g. experiences at international and national level on preventive measures, preparedness and mitigation, best practices at micro level by voluntary and nongovernmental organizations in drought prone areas for sustenance of livelihood through watershed-management projects, change of cultivation methods, use of modern irrigation methods such as drip irrigation etc.).

E. Awareness Generation: Governments, both at the level of Centre and States, should provide information on all aspects of drought to the people. State Governments should develop a drought management information system (DMIS) on different aspects of drought management. The DMIS will help in creating public awareness, with special emphasis on risk reduction, in the following ways:

- Identification of vulnerable communities.
- Sensitization programmes, and drought management educational programmes.
- Training and capacity building of communities.

- Media management.

F. Research and Development (R&D): Large-scale drought mitigation projects will have to be preceded by pilot/special studies and customised research programmes. An illustrative outline of such studies/programmes can be as under:

- Research programme in agricultural universities to study rainfall patterns and forecasting of drought at sub-district level, in the chronically drought affected areas.
- Evaluation programmes for vulnerability and impact analysis of interventions by Central and State Government in drought affected States through reputed Institutes like National Institute of Nutrition, National Institute of Rural Development etc.
- Developing and introducing drought tolerant varieties of crops.

G. Community Participation: Community participation is an essential feature of drought mitigation programmes. Rainwater harvesting holds the key to drought mitigation and the government policies emphasize community-based water resource management. Women self-help groups (SHGs) can play an important role in a large number of measures targeted at drought mitigation. SHGs can be involved in rainwater harvesting, running PDS and day care centres and overseeing water distribution and utilization in their community. They will also take measures to promote greater equity and efficiency in natural resource management. Drought relief and mitigation measures need to be, to the extent possible, implemented through the panchayati raj institutions in order to improve the efficacy of delivery.

5.5 Programmes for Treatment of Degraded Lands

With a view to prevent land degradation and to sustain agricultural production, Department of Agriculture and Cooperation of the Ministry of Agriculture, Government of India is implementing various soil and water conservation programmes for treatment/reclamation of degraded lands. The objectives of various schemes/programmes are as under:

1. Implementation of soil and water conservation programmes for the prevention of soil erosion and land degradation in order to sustain agriculture production on watershed approach basis.
2. Formulation of policies and strategies to address the issues regarding sustainable agriculture in the context of risks associated with climate change and suggesting appropriate adaptation and mitigation measures to combat adverse effect of climate change on agriculture production including natural resources.
3. Reclamation and development of land affected by alkalinity and acidity.
4. Providing scientific database on soil and land uses for watershed planning and programme implementation.
5. Prevention of runoff and premature siltation of the reservoirs, thereby sustaining irrigation potential and water storage capacity of the various reservoirs across the country.
6. Encouraging settled cultivation in lieu of shifting cultivation among the Jhumia families of the States of North-East Region and restoration of ecology and checking land degradation.
7. Creation of employment opportunities and improving the socio-economic conditions of the rural community.
8. Development of human resources through training and capacity building programmes of state officials working in the area of soil and water conservation.

The Central Sector Schemes and Centrally Sponsored Programmes being implemented by the Government are described below.

5.5.1 Soil and Land Use Survey of India (SLUSI): SLUSI is a subordinate office of the Department of Agriculture, Cooperation and Famers Welfare, established in the year 1958 with its headquarters in New Delhi. SLUSI has 7 Regional Centres located at NOIDA (Uttar Pradesh), Kolkata,

Bengaluru, Nagpur, Hyderabad, Ahmedabad and Ranchi. The major mandate of SLUSI is to conduct various types of soil and land resource surveys and provide a soil-related database for the planning and implementation of programmes relating to soil and water conservation for the development of degraded lands. The main activities of SLUSI are the following:

1. Rapid reconnaissance survey (RRS) of the catchments of river valley projects and flood prone rivers for categorization of watersheds.
2. Detailed soil survey (DSS) of selected *very high* and *high* priority watersheds to provide a sound database for execution of soil conservation measures as well as for scientific land use planning.
3. Soil resource mapping (SRM).
4. Development of GIS-based web server.
5. Organizing short training courses on soil surveys and data use for planning of watershed management.
6. Consultancy projects in soil mapping.

The organization is engaged in conducting soil surveys to provide database for land-based developmental programmes for the user organizations. Besides, SLUSI provides training to the State officials drawn from the state departments such as agriculture, soil conservation, soil survey, forestry and irrigation.

5.5.2 Soil Conservation Training Centre, Damodar Valley Corporation (DVC), Hazaribagh, Jharkhand: This Centre organizes medium and short duration training courses every year for field functionaries and project officers of the State Governments engaged in implementation of soil and water conservation programmes.

5.5.3 Soil Conservation in the Catchments of River Valley Projects and Flood Prone Rivers: Presently, this Centrally-sponsored programme is being implemented in 60 catchments of 28 States (all the States except Goa) including all the States of North-Eastern Region of the country. The total catchment area of ongoing programme is about 113.51 million hectares, out of which an area about 30.15 million hectares is categorized as a priority area, needing urgent treatment under this programme. The catchments are mostly inter-state in nature and all categories of lands viz. agriculture, waste lands and forests are taken up for treatment in an integrated manner on watershed approach basis. The main objectives are the following:

1. Prevention of land degradation by adopting a multi-disciplinary integrated approach for soil conservation and watershed management in catchment areas.
2. Improvement of land capability and moisture regime in the watersheds.
3. Promotion of land uses to match land capability.
4. Prevention of runoff and premature siltation of the reservoirs, thereby sustaining irrigation potential and water storage capacity of the various reservoirs across the country.

Under this programme, various soil and water conservation measures are implemented on watershed approach. With a view to assess the impact of soil and water conservation measures, a system of continuous monitoring of rainfall, runoff and sediment parameters prior to, during and after treatment of the watershed are collected and analyzed for assessing the impact of watershed interventions. For this purpose, an amount of 3 percent of the allocated funds are earmarked for meeting the cost of establishment of sediment monitoring stations (SMSs) and also to meet the operational cost of these SMSs.

5.5.4 Reclamation and Development of Alkali and Acid Soils (RADAS): This Programme was restructured and launched during Eleventh Five Year Plan (2007-12) for development of alkali and acid soils in the states of Arunachal Pradesh, Mizoram, Gujarat, Haryana, Punjab, Karnataka and Rajasthan.

The Programme aims at improving physical conditions and the productivity status of alkali and acid soils for restoring optimum crop production. The major components permissible under the programme are on farm development, viz. land levelling, bunding, community drainage systems, application of soil amendments, organic manures, crop/horticultural/fuel wood production etc.

5.5.5 Watershed Development Project in Shifting Cultivation Areas: This Scheme was launched during the Eighth Five Year Plan (1992-1997) in 7 states of North-Eastern Region, namely Arunachal Pradesh, Assam, Manipur, Meghalaya, Mizoram, Nagaland and Tripura with 100 percent Special Central Assistance to the State Plan. The major objectives of the scheme are the following:

1. Protection of hill slopes of Jhum areas through soil and water conservation measures on a watershed basis to reduce further land degradation.
2. Encouragement and assistance to Jhumia families to develop Jhum land for productive uses with improved cultivation and suitable package of practices leading to settled cultivation.
3. Improvement of the socio-economic status of Jhumia families through household/land based activities.
4. Mitigation of the ill effects of shifting cultivation by introducing appropriate land use methods through improved technologies.

Under this scheme, treatment of arable and non-arable land through drainage line treatment, water harvesting structures, farm ponds, horticulture, afforestation, silvipasture, crop demonstration etc. are taken up.

To sum up, drought is a normal, recurrent feature of climate and characterized in terms of its spatial extension, intensity and duration. Conditions of drought appear when rainfall is deficient in relation to the statistical multi-year average for a region, over an extended period of a season or a year, or even more. Drought is a temporary aberration unlike aridity, which is a permanent feature of climate. Drought produces wide-ranging impacts that span across many sectors of the economy and are felt far beyond the area experiencing physical drought.

Direct or primary impacts of droughts are usually associated with reduced agricultural production; depleted water levels; higher livestock mortality rates and damage to wildlife and fish habitats. When direct impacts have multiplier effect through the economy and society, they are referred to as indirect impacts in terms of reduced income for farmers and agri-business, increased prices for food and timber, unemployment, reduced purchasing capacity and demand for consumption, default on agricultural loans, and reduction in agricultural employment leading to migration etc. 971 blocks of 183 districts covering an area of about 74.6 million hectare have been identified as drought prone areas of the country. Most of the drought prone areas lie in the arid, semi-arid and sub-humid regions. States experience enormous fiscal stress in coping with drought as expenditure on relief measures often takes precedence over mitigation, placing acute burden on the State budget and thus relegating development plans.

A greater emphasis on mitigation measures can reduce incidence and severity of drought, sustain crop production and save resources spent recurrently on relief. Scientific advances in climate forecasting, information and telecommunications technology and spread of participatory democracy provide tremendous opportunities to develop an effective system for monitoring and managing drought, establishing drought forecasting systems, laying out location specific crop contingency plans, implementing timely relief programmes and focusing on long term drought proofing programmes e.g. watershed development, water harvesting etc.

Drought mitigation measures are initiatives undertaken to reduce the incidence or minimise impacts of drought. Besides drought proofing, these measures help in adapting to climate change, restoring ecological balance and bringing development benefits to the people.

However, drought mitigation programmes are not construed

to be stand-alone interventions that are implemented only in the wake of a drought; but form part of developmental planning in the domain of soil conservation, watershed development and forestry. As such, drought mitigation measures are mainstreamed in regular development programmes of Central and State Governments. Government's policy towards drought management has changed considerably over the years and now rests upon early warning and preparedness, crisis management response, medium and long-term drought mitigation measures with greater application of state-of-the art technology and scientific tools.

6

Disaster Mitigation and Management in India

6.1 India's Vulnerability to Natural Disasters

Many regions in India are highly vulnerable to natural and other disasters on account of geological conditions. About 60 percent of the landmass is susceptible to earthquakes. Over 8 percent area is prone to floods and 8,000 kilometres of coastline is prone to cyclones. Similarly, 68 percent area is susceptible to drought. All this entails huge economic losses and causes developmental setbacks. Disasters are no longer limited to natural catastrophes. Man-made emergencies often cause bigger disasters in terms of fatalities and economic losses.

The Tsunami disaster—which struck 5 coastal States/UTs of India in December 2004—has further highlighted India's vulnerability to natural disasters. Fire incidents, industrial accidents and other man-made disasters involving chemical, biological and radioactive materials are additional hazards which have underscored the need for strengthening mitigation, preparedness and response measures.

With urbanization and concentration of population in metropolitan cities, more and more people are becoming vulnerable to locational disasters. Therefore, the development process needs to be sensitive towards disaster prevention, preparedness and mitigation. In fact, disaster management has emerged as a high priority area for the country. Going beyond the historical focus on relief and rehabilitation after the event, there is a need to look ahead and plan for disaster preparedness and mitigation in order to ensure that periodic shocks to development efforts are minimized. Disaster-risk reduction has not been highlighted in the policies and programmes of various plan schemes. India's commitment to mainstreaming disaster-risk reduction into the process of development planning at all levels, so as to achieve sustainable development, is yet to be carried forward across sectors through actionable programmes for achieving the desired result.

6.2 Disaster Management and Development Planning

Mainstreaming disaster management into the development planning process essentially means looking critically at each activity that is being planned, not only from the perspective of reducing the disaster vulnerability of that activity, but also from the perspective of minimizing that activity's potential contribution to the hazard. Every development plan of a Ministry/Department should incorporate elements of impact assessment, risk reduction and the do-no-harm approach. Examples of this approach are urban planning and zoning, upgradation of building codes and their effective enforcement, adoption of disaster resilient housing designs and construction of school and hospitals, flood proofing, response preparedness planning, insurance, establishment of early warning systems for various types of disasters, generating community awareness, creating technical competence and promoting research among engineers, architects, health experts and scientists.

State Governments need to give priority in their plans for schemes regarding hazard identification and risk assessment once they have prepared the project paper, completed preliminary work and drawn up the details of the scheme etc. Hazard identification and risk assessment across the country must be bound by uniformly followed procedures, fine-tuned to local conditions. In the absence of such procedures, any sporadic activity based on some ad hoc procedure carries the potential of doing more harm than good. It is essential that while clearing plans of State Governments, there is an integrated approach particularly for the creation of adequate capacity for relief and rescue operations.

Financing of disaster relief is an important aspect of

federal fiscal relations. There are significant variations in the disaster proneness profiles of different states and wide regional disparities in terms of levels of economic development. This implies that the coping capacity of a majority of the states to deal with disasters on their own is inadequate. This is compounded by the fact that the poorer states are often the most disaster prone. The financing of disaster relief has, as a result, come to be firmly accepted as a joint endeavour of the Central and State Governments.

Outside the framework of plan schemes, many innovative measures can be adopted to encourage disaster risk reduction measures by the corporate sector, non-governmental organizations and individuals. Fiscal measures like rebates on income and property tax for retrofitting unsafe buildings, compulsory risk insurance for bank loan on all types of properties etc. shall definitely help to mobilize resources for safe construction and retrofitting of existing constructions in all disaster prone areas. Similarly, many innovative measures may be taken for promoting public-private-community partnership for disaster risk reduction.

6.3 Disaster Management Act, 2005

This legislation was enacted for establishing requisite institutional mechanisms for drawing up and monitoring the implementation of disaster management plans, ensuring measures by various wings of Government for prevention and mitigating effects of disasters, and for undertaking a holistic, co-ordinated and prompt response to any disaster situation.

The National Policy on Disaster Management (NPDM) issued under the Disaster Management Act envisages building a safe and disaster-resilient India by developing a holistic, proactive, multiple disaster-oriented and technology-driven strategy through a culture of prevention, mitigation, preparedness and response. Under the provisions of the Act, the National Disaster Management Authority (NDMA) has been set up and the National Disaster Response Force raised.

The Disaster Management Act provides for the effective management of disasters and all related matters, including the mechanisms for funding disaster relief and response. The Act defines disaster in very broad terms to include both natural and man-made disasters. It envisages the constitution of two types of funds, one for disaster response and the other for mitigation. These are to be set up at the national, state and district levels.

Thus, for disaster response, the Act envisages a National Disaster Response Fund (NDRF), a State Disaster Response Fund (SDRF) in each State and, within the States, a District Disaster Response Fund (DDRF) in each district. Similarly, the Act envisages a National Disaster Mitigation Fund (NDMF), State Disaster Mitigation Funds (SDMF) and District Disaster Mitigation Funds (DDMF) for disaster mitigation. So far, at the national level, only the NDRF has been constituted. All State Governments have constituted an SDRF, but only a few have constituted an SDMF. Very few State Governments have constituted District Disaster Response Funds (DDRFs). During a disaster, relief activities at the district level are generally carried out through transfers from the SDRF.

6.3.1 National Disaster Management Authority (NDMA): NDMA under the Chairmanship of the Prime Minister was set up in terms of the Disaster Management Act, 2005. NDMA is an apex body responsible for laying down of policies, plans and guidelines on disaster management so as to ensure timely and effective response to disaster. NDMA has prepared a Draft National Policy on Disaster Management and has taken up preparation of guidelines on prevention, mitigation, response and recovery in regard to various types of disasters such as earthquake, flood, landslides, industrial disaster etc. The guidelines on management of earthquake, chemical disasters and chemical (industrial) disasters have also been circulated.

NDMA, in consultation with National Informatics Centre (NIC), National Spatial Data Infrastructure (NSDI), National Remote Sensing Agency (NRSA) and Department of Space is actively promoting the use of GIS in all institutions concerned with disaster management. Priority will have to be given to highly vulnerable areas for generating database for mapping as well as hazard vulnerability and risk assessment related to disasters like cyclone, flood earthquake, chemical, industrial etc.

6.3.2 State and District Disaster Management Authorities (SDMAs): The State Governments are in the process of setting up SDMAs. The provisions of the Disaster Management Act, 2005 relevant to the States/UTs were brought into force w.e.f. August 1, 2007. While Arunachal Pradesh, Goa, Gujarat, Himachal Pradesh, Kerala, Mizoram, Puducherry, Punjab and Uttar Pradesh are reported to have constituted SDMAs as per the Act, the other States/UTs are in the process of constituting the same. Existing institutions have been strengthened to give priority to, and enhance their capacities for, rescue, relief, rehabilitation and mitigation. Some States have raised a State Disaster Response Force along the lines of the National Disaster Response Force. States have also focussed on training and generating awareness to help cope with disasters.

6.4 National Disaster Response Force (NDRF)

A 12 battalion strong NDRF has been set up comprising of 216 specialized response teams on various types of disasters, including the ones for nuclear, biological and chemical (NBC) disasters. The civil defence set up in the country is also proposed to be revamped to strengthen the local efforts for disaster preparedness and effective response. Similarly, the fire services are also proposed to be strengthened/modernized to convert them into multi-hazard response force.

The total strength of each battalion is 1,149 and they have been equipped and trained to respond to natural as well as man-made disasters. The NDRF battalions are placed in 12 different locations across the country based on the vulnerability profile of the areas.

6.5 National Institute of Disaster Management (NIDM)

NIDM has been set up for training, capacity building, research and documentation on various natural and man-made disasters. A comprehensive human resource plan for disaster management has also been developed. Disaster management has been included in the curriculum of middle and secondary school education. The subject has also been included in the post-induction and in-service training of civil and police officers. Modules have also been identified to include disaster management aspects in the course curriculum for engineering, architecture and medical degrees. A committee of experts has finalized the model building bye laws for town and country planning legislations, land use zonation and development control legislations. The municipalities and city development authorities all over the country have been advised to make necessary changes in their respective bye laws and regulations in accordance with the model laws.

Bureau of Indian Standards (BIS) has issued building codes for construction of different types of buildings in different seismic zones in the country. The National Building Code has also been revised taking into consideration the natural hazards and risks of various regions of the country.

6.6 Twelfth Five Year Plan (2012-17) on Disaster Management

According to the Twelfth Plan, "A development strategy under the planning process has risk management as one of its key components. Globally, there is an increasing recognition that disasters affect growth and the poorer sections of society gets a major share of the impact. Therefore, there is a consensus that investing in prevention and mitigation is economically and socially more beneficial than expenditure in relief and rehabilitation. In a recent World Bank Study, it has been established that one dollar spent on prevention is ten times more valuable than a dollar spent on relief in net present value.

The large size of the Indian continent, varied geography, national features, climate, and effects of economic development and growth process results in number of risks. These are clearly both due to natural hazards and effect of human development process on nature. The human and economic losses from disaster are economically high in the country as compared to the other developed nations of the world. According to a World Bank study titled 'National Hazards, Unnatural Disasters', India faces losses up to 2 percent of its GDP due to natural disasters. Disasters impact growth particularly of the poor and vulnerable sections of society.

The strategy in the initial years of the planning process has been to handle only some of the natural hazards, like floods and water proofing against drought. It has, however, been gradually realised that a more comprehensive approach is called for. This must lead to mitigation of ill-effects of disasters, and a development process which encompasses within itself, a strategy for mitigation of human misery and adverse consequences of the disaster.

The Tenth Five Year Plan initiated the process of shift from relief and response centric disaster management to prevention, mitigation, and preparedness as means to revert or more effective handling of the disasters. The Plan prescribed a strategy of disaster management which included three important components: First, policy guidelines on preparation of developmental plans across sectors for integrating disaster management into developmental plans and a specific scheme.

Second, a multi-pronged strategy for the total risk management. Third, recognition of a need for plan expenditure on disaster management and preventive measures in addition to calamity relief fund. In pursuance of this strategy, the National Disaster Management Authority was proposed under National Disaster Management Act 2005. The process of strengthening disaster preparedness was further strengthened in the Eleventh Plan. The disaster management authority was set up at the national level. It formulated extensive guidelines and a national policy on disaster management. Similar to above, state and district level authorities were also set up gradually". [1]

6.7 Disaster Management and Finance Commissions

The financing of disaster relief has been an important aspect of federal fiscal relations. Successive Finance Commissions since the Second Finance Commission have provided *margin money* for such contingencies. From the time of the Sixth Finance Commission, the terms of reference (ToR) have specifically entrusted each Commission with the task of considering the arrangements between the Union Government and State Governments relating to financing disaster management and making recommendations on the subject. When earlier Commissions examined the issues, there was no specific law relating to disaster management. By the time the Thirteenth Finance Commission was constituted, Parliament had enacted the Disaster Management Act, 2005.

The primary responsibility for undertaking rescue, relief and rehabilitation measures during a disaster lies with the State Governments. The Union Government supplements their efforts through logistic and financial support during severe natural disasters. Based on the recommendations of the Thirteenth Finance Commission, the National Calamity Contingency Fund (NCCF) was merged with the newly constituted NDRF. Similarly, at the state level, the Calamity Relief Fund (CRF) was merged with the SDRF. Financing of the SDRF is based on the recommendations of the Finance Commissions, which determine the annual size of the Funds as well as the respective contributions of the Union and State Governments.

The NDRF is financed through the levy of a cess on certain items, chargeable to excise and customs duty, and approved annually through the Finance Bill. The requirement for funds beyond what is available under the NDRF is met through general budgetary resources. The Union Government has issued guidelines on the items and norms of assistance, the natural calamities eligible for funding, accounting norms and audit arrangements, procedures for the release of funds and the respective share of each State as well as its contribution under the award of the Thirteenth Finance Commission towards the SDRF.

The notified calamities based on the recommendations of successive Finance Commissions include cyclones, droughts, earthquakes, fires, floods, tsunamis, hailstorms, landslides, avalanches, cloud bursts and pest attacks. The Union Government has recently added 'cold waves' and 'frost' to that list. The norms under each type of disaster were last revised in November 2013.

6.8 Gujarat Institute of Disaster Management (GIDM)

GIDM registered as an autonomous body on 10.07.2012 under the provisions of the Societies Registration Act, 1860 and Bombay Public Trust Act, 1950 to allow and function "GIDM" as an autonomous body to carryout various functions for implementation of Disaster Related Programmes on training and capacity building of the stakeholders.

Since April 1, 2004, Gujarat Institute of Disaster Management was functioning in the premises of the Sardar Patel Institute of Public Administration (SPIPA), Ahmedabad but under the administrative control of GSDMA. In August 2012, Gujarat Institute of Disaster Management shifted to its newly constructed building at Raisan, Gandhinagar. The Institute has trained Government officials, PRI members, personnel from NGOs etc. by conducting training programmes in different aspects of Disaster Management.

The GIDM was established as the research and training wing of the Gujarat State Disaster Management Authority (GSDMA) on 26th January 2004 by the Government of Gujarat with the aim of Human Resource Development in the state. With state of the art facilities, the main thrust of the institute is to provide proper insight, knowledge and skills to stackholders at various levels in the state, so that they can accomplish appropriate disaster management tasks. GIDM will focus upon enriching the capacities (specific areas of prevention, mitigation, preparedness, relief, recovery, rehabilitation, reconstruction and sustainable development) of the officials and other functionaries of the state and district administration for response to emergencies and in the longer run, it will take up similar capacity building for other stakeholders. To make available to all stakeholders, the knowledge and skills required to develop their individual competencies, gain appropriate understanding and accomplish appropriate disaster management tasks, while establishing and strengthening overall disaster.

The National Disaster Management (NDM) Cell of GIDM was providing Disaster Management training to Government Officers in the state. Further the Govt. of Gujarat in Revenue Department, declared the erstwhile "Disaster Management Cell" as "Gujarat Institute of Disaster Management".

The World Bank included the project of establishing Gujarat Institute of Disaster Management as part of Phase-II of the Gujarat Emergency Earthquake Reconstruction Project (GEERP). It was for the first time that the international funding agency included the establishment of a premier permanent institute which deals with training related to disaster management in any such reconstruction projects.

The Government of Gujarat allotted land for establishing the Institute at Village Raisan near Koba Gandhinagar and the construction work has been completed. The Authority has also made yearly provision of ₹ 10 crore in the State budget of the year 2012-13 and subsequent years under the head 101 Natural Calamities—Assisted to Disaster Management Authority—Gujarat Institute of Disaster Management (GIDM). Presently, funds are available with GSDMA.

6.8.1 Vision and Mission of GIDM: Make available to all stakeholders the knowledge and skills required to develop their individual competencies, gain appropriate understanding and accomplish appropriate disaster management tasks, while establishing and strengthening overall disaster management institutional capacities at the state level and below.

6.8.2 Objective of GIDM:

1. To serve as the apex institute in the state for disaster

management capacity building.

2. To provide disaster related training to all the stakeholders.
3. To assess the training needs of the different government departments, offices, boards, corporations, local bodies and non-governmental organizations and others who may consent to be associated with it.
4. To undertake activities for human resource development, public education and community awareness, safety etc. in disaster education and management.
5. To undertake quality research projects on disaster management and mitigation covering both natural and human-induced disasters.
6. To act as a resource centre and clearing house of information on disaster management by documentation of field experiences including case studies, lessons learnt and best practices.
7. To establish and maintain libraries and spread knowledge of disaster management.
8. To facilitate partnerships with willing eminent national and international organizations, universities, institutions, bodies and individuals specialized in disaster management.
9. To serve as a centre of excellence on disaster mitigation and management.
10. To arrange for financial assistance, to raise money through financial instruments, bonds, deposits etc. permissible under the provisions of Societies, Registration Act, 1860 and the Bombay Public Trust Act, 1950 and management, administration, investment and reinvestment of funds so raised or available with the institute.
11. To run and award degree, diploma, certificate courses on disaster management at its own or with the affiliation of any other institutes and universities.
12. To undertake publication of newsletters, journals, reports, occasional papers, study reports etc.

To sum up, disaster preparedness was in the past viewed only in terms of relief and rehabilitation after the event took place. Current thinking realizes that there is much more to disaster management, especially strategies which need to be in place and measures taken before the event to mitigate the effects of disasters.

Endnote

1. Government of India, Planning Commission, *Twelfth Five Year Plan* (2012-17), Volume I, Chapter 10, paras 10.65 to 10.68.

7

Disaster Management Education

7.1 Role of Education for Disaster Preparedness

Education has been considered as a potential instrument for achieving people's participation in environmental management in general and disaster mitigation in particular. In India, education for disaster management is a trans-disciplinary exercise aimed at developing knowledge, skills and values at all levels. Government of India in its Tenth Five Year Plan integrated this management in the educational sector to enhance knowledge, skills and values to reduce the impact of disasters through public awareness. Education for disaster management has provided the awareness of protective actions for different hazards. Thus, a group of people and communities learn how to save themselves from the entire crisis. The prime theme of this education is to design the awareness programme among the public in both short-term and long-term preparedness procedures and to furnish the knowledge—how to prepare in advance, how to act on warnings and how to reduce the risks from their home and communities.

Disaster education (DE) plays an important role in both: (a) minimizing the impacts of the disaster on communities, and (b) improving disaster preparedness which helps people to restore normalcy and provide them physical and psychological assistance after such a traumatic experience. The World Bank Good Practice note elucidates that disaster education pursues the dual objectives of promoting short-term recovery and ensuring better disaster preparedness over the medium-term. Disaster education ought to be for all people, men or women, young or old, literate or illiterate and it must address their specific needs as individuals and groups.

Hence, we need a breed of experts and specialists in all the diverse aspects of disaster mitigation and management for which enough space can be found in our universities and higher institutes of learning; a much broader framework alone can meet the demand for disaster education for all. Disaster education has now become a bridge between research and development (especially on the issues leading to sustainable protection of natural and built environment) at one end and solving of real life problems (consultancy or practice) on the other end.

Disaster education initiates preparedness activities and provides knowledge about the hazard and its risk and helps to overcome the bias in risk perception and makes people perceive the risk with clarity and certainty. The knowledge about disasters and its management is not only vital for the government officials, paramilitary forces, health care practitioners and other stakeholders but also for the entire population. If common people know what has to be done when a tragedy strikes and can take immediate measures, damage to life and property can be substantially reduced.

Everyone should know what disasters are, why they happen, how they happen, what are its consequences, and what steps should be taken before, during and after a disaster strikes. The basic purpose of this education is to enhance the managerial skills of paramilitary forces and others involved in handling disasters. The coordination between different government agencies is very important at the time of disaster. The programme focuses on prevention, preparedness, crisis management, decision-making, relief and rehabilitation, and ensuring synergy between various organizations involved in disaster management.

Knowledge and experience of disasters empower the people to prepare themselves in a better fashion. Moreover, disaster education programmes enhance knowledge and awareness of the people for appropriate protective actions to be taken to reduce damages to life and property as a result of disasters. People having this education would able to prepare their families better than others.

In the current era, disaster education is a new area which is still not fully implemented all over the world. But, according to reports, "climate change is occurring in dramatic ways; the sea level is likely to rise up to a meter by mid of this century and CO_2 emissions too would have a major impact on the environment". Thus, it is obligatory to have this particular education for safeguard of our environment and by getting this knowledge, people should learn how to recover or rescue their families, households and communities. Disaster management knowledge otherwise also means do's or don'ts in a crisis situation.

Education in hazard mitigation and preparedness expands to include interdisciplinary and holistic degree programmes. Members of higher education community will have to invent university based programmes that move away from traditional disciplines towards interdisciplinary education that solves the 'real world' problems entailed in linking hazards and sustainability. Thus, educated people are following scientific and modern methods of disaster preparedness, whereas due to lack of this education, people in rural areas are still following the traditional methods.

According to G.N. Ritchie, "disaster can be reduced by education, through proper information and public awareness". People should play a crucial role in reducing disaster effects with participation by government and non-governmental organizations or the community itself should aware and train people and fulfil their responsibilities. Programmes for education, public awareness, information and training conducted at the community level, increasing public awareness to become self-reliant and self-help should be regularly organised for adequate preparedness and response whenever a disaster strikes.

According to Maury, "preparedness for disaster risk reduction is not possible through short-term measures during pre-disaster phase but because of appropriate disaster legislations, operational planning, education and training, and other long-term responses for its preparedness". People should better prepare themselves through efficient leadership, readiness, support, and assistance in pre-empting the consequences of disasters, mitigating the suffering they bring forth, organizing effective response and launching appropriate recovery measures.

7.2 Disaster Management Education in India

The term disaster management education, as used herein, "refers to the transfer of knowledge and the facilitation of understanding regarding the management of hazards and disasters; specifically, the recognition of natural and human-induced hazards and reduction of vulnerability" (International Journal of Mass Emergencies and Disasters). India as a nation is prone to various hazards due to its unique geo-climatic conditions as a result of which floods, droughts, cyclones, earthquakes and landslides have been occurring consistently. About 60 percent of the land mass is prone to earthquakes of various intensities; over 40 million hectares is prone to floods; about 8 percent of the total area is prone to cyclones and 68 percent of the area is susceptible to droughts. Hence, experience and knowledge of past shows that when attention has been given to adequate preparedness measures, the loss of life and property has considerably reduced.

However, incorporating disaster education in educational institutions either in formal school curriculum or in non-formal education makes the community and its members more efficient for disaster preparedness. Government of India, Ministry of Human Resource Development (MoHRD) has emphasized the need for integrating disaster education in the existing education system in India. The first attempt has been taken by the Central Board of Secondary Education (CBSE), to integrate a short course on disaster management in the school curriculum.

The curriculum on disaster management includes the

following topics:

1. The nature and types of hazards.
2. Natural and man-made disasters and their management.
3. Efforts made in various regions for disaster preparedness and mitigation.
4. Role of community and schools in disaster management.
5. Partnership with various governmental and non-governmental agencies.
6. Use of modern and scientific technologies to combat disasters.
7. Survival skills.

Disaster management education in the pedagogic institutes intends to cross the boundaries of the curriculum and classrooms and make learning local-specific, involving families and communities. The Board strongly believes that "educating a child is educating a community". Children are the dynamic and powerful force of change and are supporters in creating awareness in the community and contribute in a unique manner with their energy and vision to find local solutions. After the recommendations by the Ministry of Home Affairs, Government of India, various state governments are in the process of integrating disaster management in school education.

1 **Tamil Nadu:** The course curriculum has been drafted and will be incorporated soon.

2 **Odisha:** *Disaster Risk Safety* was launched by Orissa State Disaster Management Authority (OSDMA) for students and it integrates a chapter on disaster management in the class VII geography syllabus.

3 **Maharashtra:** The education department has initiated a process of incorporating disaster management in school curriculum.

4 **Bihar:** Disaster management has been carried forward through *Sarva Shiksha Abhiyan* and disaster management in course curriculum has been incorporated from class V onwards in social sciences.

5 **Gujarat:** Disaster management classes have been initiated for classes VII, VIII and IX.
6 **West Bengal:** State authorities are in the process of introducing disaster management course in the schools run by civic bodies.
7 **Jharkhand:** The state government has included a text book on disaster management in the class IX syllabus as part of the social science subject and Xavier Institute of Social Service has also incorporated four papers in their business management course.

7.3 National Disaster Management Training and Education for Disaster Preparedness

Proper training and relevant education gives a clear and coherent picture while dealing with disasters. Training is a vital component for disaster recovery designs for varied target groups who will be providing personal, psycho-social and community-based services to the affected communities. Information and training on ways for better response to mitigate disaster effects go a long way in building the capacity and resilience of the country (India Disaster Knowledge Network). Training is an integral part of capacity building as trained personnel respond much better to different disasters and appreciate the need for preventive measures. The directions in this regard are as under:

1. The multi-sectoral and multi-hazard prevention based approach to disaster management requires specific professional inputs. Professional training in disaster management should be built in the existing pedagogic research and education. Specialized courses for disaster management may be developed by universities and professional teaching institutions, and it should be treated as a distinct academic and professional discipline, something that the American education system has done successfully. In addition to separate diploma/degree courses in disaster management, the subject needs to be

discussed and taught as a specific component in professional and specialized courses like medicine, nursing, engineering, environmental sciences, architecture, and town and country planning.

2. The focus towards preventive disaster management and development of a national ethos of prevention, calls for an awareness generation at all levels. An appropriate component of disaster awareness at the school level will help increase awareness among children and, in many cases, parents and other family members through their children. Curriculum development with a focus towards dissemination of disaster related information on a sustained basis covering junior, middle and high schools may be worked out by the different school boards in the country.
3. Training facilities for government personnel involved in disaster management are conducted at the national level by the National Centre for Disaster Management (NCDM) at the Indian Institute of Public Administration (IIPA), in New Delhi which functions as the nodal institution in the country for training, research and documentation of disasters. At the State level, disaster management cells operating within the state Administrative Training Institutes (ATIs) provide the necessary training. Presently, 24 ATIs have dedicated faculties. There is a need for strengthening specialized training including training of personnel in disaster response.
4. Capacity building should not be limited to professionals and personnel involved in disaster management but should also focus on building the knowledge, attitude and skills of a community to cope with the effects of disasters. Identification and training of volunteers from the community towards first response measures as well as mitigation measures is an urgent imperative. A programme of periodic drills should be introduced in vulnerable areas to enable prompt and appropriate community response in the event of a disaster,

which can help to save valuable lives.

7.4 Public Education, Training, Rehearsal and Experience for Disaster Preparedness

Public education comprises of strategies, plans and guidelines which form the foundation of long-term public information on how to act in the event of occurrence of a disaster. Public education has particularly shown strong usage with regards to educating for public safety. Educational campaigns have been focused on topics including road safety, toxic chemicals, safe sex, skin cancer, health safety, fire prevention and response, what to do at the scene of an accident, sudden infant death syndrome prevention (IDSP) and many more. Disaster public education planning and implementation circle shows how public education helps in handling hazards from one stage to another.

Disaster public education is a key component to a jurisdiction's emergency management programme, and it includes goals, objectives and tasks not directly related to a single incident, and response and recovery operation. Planning for public education should also be closely linked to the public information functions, which is frequently tested through exercises and actual events. Although messages and delivery techniques differ, jurisdiction's public education programmers and preparations for disseminating information to the public during a crisis should benefit from the hazard identification process and knowledge of populations, including locations of people with special need.

The survey conducted in September 2005 by the *Washington Post*, Kaiser Family Foundation and Harvard University highlights the needs for better public education and outreach pre-disaster. The survey was conducted among 680 adults who were evacuated from the Gulf Coast of Houston following Hurricane Katrina.

The study reveals that after getting knowledge about disaster preparedness, merely 80 percent people were able to evacuate their rooms before the storm hit, whereas the rest 20 percent were either physically inefficient or they did not want to leave their place as they were closely attached with that place. Following are some of the key areas for strengthening public education:

1 Establishing a national framework for public education that includes a single national vision and clear connection among locals, state and nation as a whole. Build up a national vision that recognizes the importance of public education and individual understanding of disaster preparedness.

2 Public disaster education aims to bridge the gaps between sociological and psychological research on public education policy and implementation in the field of disaster. There is a significant disconnection between emergency management practice and social science research. Informative studies often become entrenched in academic circles; systems are not established for communicating findings to practitioners. Likewise, public officials may not know how or do not have time to transform academic resources and findings into practice.

3 Developing standards and hosting workshops on disaster public education can create better awareness among the people. The components of disaster public education programme outlined above should be added to the Emergency Management Accreditation Programme (EMAP) which will stand as a supplementary guide for further developmental steps. Furthermore, the content should form the basis for regional or state-level workshops on developing disaster public education programmes.

4 Education and experience mutually help in risk perception which ultimately leads to disaster preparedness as the upshot.

7.5 Education, Experience and Risk Perception for Disaster Preparedness

One model with direct paths deals with the impact of disaster experience and education on preparedness behaviour, and another with indirect paths, tested the impact of disaster experience and education through risk perception on preparedness behaviour. Thus, those people who are following the direct paths were better in their preparations than those following the indirect path.

Disaster preparedness must be supported by public education campaigning, training of response teams and rehearsals for emergency response scenarios. The aim of public awareness and education programmes is to promote an informed, alert and self-reliant community, capable of playing its full part in support of, and in co-operation with government officials. An essential part of a disaster preparedness plan is education for those who may be threatened by a disaster. Although television, radio and print media will never replace the impact of direct instructions, at the same time sensitively designed and projected messages can provide a useful supplement to the overall process.

As the preparedness plan is being developed, and upon its completion, it is important to rehearse its major elements. Rehearsals invariably expose gaps that otherwise remain overlooked. Rehearsals are most effective when they are system wide and engage as many of the disaster response players as possible. Rehearsals also keep the plans fresh during extended periods of time when no disaster strikes. Rehearsals might simulate search and rescue operations, first aid provisions, response or needs assessment, coordination meetings between major organizational players and population leaders, relief transport and logistics, and many other aspects of an emergency response.

8

Concept and Benefits of Corporate Social Responsibility (CSR)

8.1 What is CSR?

Corporate social responsibility (also called corporate conscience, corporate citizenship, social performance, sustainable responsible business) is a form of corporate self-regulation integrated into a business model. CSR policy, functions as a built-in, self-regulating mechanism whereby a business monitors and ensures its active compliance with the spirit of law, ethical standards, and international norms. In some models, a firm's implementation of CSR goes beyond compliance and engages in "actions that appear to further some social good, beyond the interests of the firm and that which is required by law". CSR is a process with an aim to embrace responsibility for the company's actions and encourage a positive impact through its activities on environment, consumers, employees, communities, stakeholders and all other members in the public sphere who may also be considered as stakeholders.

CSR is a concept, accepted the world over, whereby corporates integrate social and environmental concerns in their business operations and in interactions with their stakeholders on a voluntary basis. The three keys to an effective CSR are commitment, clarity and congruence with corporate values. Clarity is all important because each corporate entity has to frame its CSR initiatives based on the environment of the activities being carried out. Congruence is about ensuring the corporate attitude of its responsibilities towards the society. CSR should also be consistent with own values and culture of the company.

CSR encompasses a wide spectrum of issues ranging from business ethics, corporate governance and socially responsible investing (SRI) to environmental sustainability and community investment. CSR is about the behaviour of businesses over and above what is ordinarily required by regulatory bodies and legal requirements.

Presently, there is an increasing awareness about CSR the world over. Consequently, there is a concerted effort among all types of organizations to ensure that sustainable development is not lost sight of, in the pursuit of their respective goals—profit-making, social service, philanthropy etc. CSR entails the integration of social and environmental concerns by companies in their business operations as also in interactions with their stakeholders.

8.2 Main Features of CSR

These are as under:

- CSR is behaviour by businesses over and above legal requirements, voluntarily adopted because businesses deem it to be in their long-term interest.
- CSR is intrinsically linked to the concept of sustainable development. Businesses need to integrate economic, social and environmental impact in their operations.
- CSR is not an optional *add-on* to core activities of business but about the way in which business is managed.
- CSR is the alignment of business operations with social values. It takes into account the interests of stakeholders in the company's business policies and actions.
- CSR focuses on the social, environmental, and financial success of a company—the so-called *triple bottom line*—with the aim to achieve social development while achieving business success.

8.3 Origins of CSR

The concept of corporate social responsibility was formally coined in 1953 by Howard Bowen in his publication *Social Responsibilities of Businessmen*. Bowen defined CSR as the obligations of businessmen to pursue those policies, to make those decisions, or to follow those lines of action which are desirable in terms of the objectives and values of society. The concept received academic attention in 1980s and 1990s and became an issue of debate and discussion among the academic circles.

The term *corporate social responsibility* came into common use in the late 1960s and early 1970s after many multinational corporations formed the term stakeholder, meaning those on whom an organization's activities have an impact. It was used to describe corporate owners beyond shareholders as a result of an influential book by R. Edward Freeman titled, *Strategic Management: A Stakeholder Approach* in 1984.

The first company to implement CSR was Shell in 1998. Although, the literature on CSR has been developed in all countries, its practical application is more profuse in the developed countries, mainly the US and some European nations. The concept of social responsibility has been interpreted by different authors from different corporate perspectives. A number of concepts and issues are subsumed under the heading of CSR, including human rights, environmental perspective, diversity management, environmental sustainability and philanthropy.

Thus, it can be said that the concept of CSR is complex in nature. Numerous studies have investigated the link between CSR and financial performance through a theoretical as well as an empirical lens. In particular, research rooted in neo-classical economics argued that CSR unnecessarily raises a firm's costs, putting the firm in a position of competitive disadvantage vis-à-vis its competitors. Predominantly based on agency theory, some studies have argued that employing valuable resources of a firm to engage in CSR results in significant managerial benefits rather than financial benefits to the firm's shareholders.

8.4 Approaches to CSR

Many global frameworks and standards for CSR have been developed that are meant to help companies conform to principles and practices that go beyond compliance with the law.

CSR is titled to aid an organization's mission as well as a guide to what the company stands for and will uphold to its consumers. Development of business ethics is one of the forms of applied ethics that examines ethical principles and moral or ethical problems that can arise in a business environment. ISO 26000 is the recognized international standard for CSR.

A more common approach to CSR is corporate philanthropy. This includes monetary donations and aid given to local and non-local non-profit organizations and communities, including donations in areas such as arts, education, housing, health, social welfare, and the environment, among others, but excluding political contributions and commercial sponsorship of events. Some organizations do not like a philanthropy-based approach as it might not help build the skills of the local population, whereas community-based development generally leads to a more sustainable development.

Another approach is garnering increasing corporate responsibility interest. This is called creating shared value (CSV). The shared value model is based on the idea that corporate success and social welfare are interdependent. A business needs a healthy and educated workforce, and sustainable resources to compete effectively. For society to thrive, profitable and competitive businesses must be developed and supported to create income, wealth, tax revenues, and opportunities for philanthropy.

Various approaches to CSR pit businesses against society, emphasizing the costs and limitations of compliance with externally imposed social and environmental standards. CSV acknowledges trade-offs between short-term profitability and social or environmental goals, but focuses more on the opportunities for competitive advantage from building a social value proposition into corporate strategy. CSV has a limitation in that it gives the impression that only two stakeholders are important—shareholders and consumers—and belies the multi-stakeholder approach of most CSR advocates.

Many companies use the strategy of benchmarking to

compete within their respective industries in CSR policy, implementation, and effectiveness. Benchmarking involves reviewing CSR initiatives of the competitor, as well as measuring and evaluating the impact that those policies have on society and the environment, and how customers perceive competitor's CSR strategy. After a comprehensive study of competitor's strategy and an internal policy review, a comparison can be drawn and a strategy developed for competition with CSR initiatives.

8.4.1 Cost-benefit Analysis with a Resource-based View: In competitive markets, the cost-benefit analysis regarding positive financial outcomes upon implementing a CSR-based strategy, can be examined with a lens of the resource-based-view (RBV) of sustainable competitive advantage. A firm can conduct a cost-benefit analysis through a RBV-based lens to determine the optimal and appropriate level of investment in CSR, as it would with any other investments. A firm introducing a CSR-based strategy might only sustain high returns on their investment if their CSR-based strategy were inimitable by their competitors. In competitive markets, a firm introducing a CSR-based strategy might only sustain high returns on their investment and there may only be a short-lived strategic competitive advantage to implementing CSR as their competitors may adopt similar strategies.

There is, however, a long-term advantage in that competitors may also imitate CSR-based strategies in a socially responsible way. Even if a firm chooses CSR for strategic financial gains, the firm is also acting responsibly. Attention to CSR as an element in corporate strategy leads to examining CSR activities through the lens of the resource-based-view (RBV) of the firm. RBV presumes that firms are bundles of heterogeneous resources and capabilities that are imperfectly mobile across firms. Accordingly, the imperfect mobility of heterogeneous resources can result in competitive advantages for firms that have superior resources or capabilities.

8.4.2 Social Accounting, Auditing, and Reporting: Social accounting emphasizes the notion of corporate accountability. It is an approach to reporting a firm's activities which stresses the need for the identification of socially relevant behaviour, the determination of those to whom the company is accountable for its social performance and the development of appropriate measures and reporting techniques.

8.4.3 Social License: Social license generally refers to a local community's acceptance or approval of a company's project or ongoing presence in an area. It is increasingly recognized by various stakeholders and communities as a prerequisite to development. The development of social license occurs outside of formal permitting or regulatory processes, and requires sustained investment by proponents to acquire and maintain social capital within the context of trust-based relationships. Often intangible and informal, social license can nevertheless be realized through a robust suite of actions cantered on timely and effective communication, meaningful dialogue, and ethical and responsible behaviour.

Local conditions, needs, and customs vary considerably and are often opaque, but have a significant impact on the likely success of various approaches to building social capital and trust. These regional and cultural differences demand a flexible and responsive approach and must be understood early in order to enable the development and implementation of an effective strategy to earn and maintain social license. Governments could facilitate the necessary stakeholder mapping in regions for which they are responsible and provide a regulatory framework that sets companies on the right path for engagement with communities and stakeholders.

Social media tools empower stakeholders and communities to access and share information on company behaviours, technologies, and projects as they are implemented around the world. Understanding and managing this reality will be important for companies seeking social license. Voluntary measures integral to corporate-responsibility frameworks contribute to achieving social license, particularly by enhancing a company's reputation and strengthening of its capacity for effective communication, engagement, and collaboration. However, such measures do not obviate the need for project-specific action to earn and maintain social license.

8.5 Principle of Triple Bottom Line (TBL)

People, planet, profit (also known as the *triple bottom line*) are words that should be used and practiced in every move an organization makes. People relate to fair and beneficial business practices towards labour, the community and region where corporation conducts its business. Planet refers to sustainable environmental practices. A triple bottom line company does not produce harmful or destructive products such as weapons, toxic chemicals or batteries containing dangerous heavy metals. Profit is the economic value created by the organization after deducting the cost of all inputs, including the cost of the capital tied up. It, therefore, differs from traditional accounting definitions of profit.

The main principles involving corporate social responsibility involve economic, legal, ethical and discretionary aspects. A corporation needs to generate profits, while operating within the laws of the state. The corporation also needs to be ethical, but has the right to be discretional about the decisions it makes. Levels of corporate social responsiveness to an issue include being reactive, defensive, responsive and interactive. All these terms are useful in management issues. Selecting when and how to act can make a difference in the outcome of the action taken.

Increasingly, corporations are motivated to become more socially responsible because their most important stakeholders expect them to understand and address the social and community issues that are relevant to them. Understanding what causes are important to employees is usually the first priority because of the many interrelated business benefits that can be derived from increased employee engagement (i.e. more loyalty, improved recruitment, increased retention, higher productivity, and so on). Key external stakeholders include customers, investors (particularly institutional investors), communities in the areas where the corporation operates its facilities, regulators, academics, and the media.

8.6 Potential Business Benefits of CSR

The definition of CSR used within an organization can vary from the strict *stakeholder impacts* definition used by many CSR advocates and will often include charitable efforts and volunteering. CSR may be based within the human resources, business development or public relations departments of an organisation, or may be given a separate unit reporting to the CEO or in some cases directly to the board. Some companies may implement CSR-type values without a clearly defined team or programme.

The scale and nature of the benefits of CSR for an organization can vary depending on the nature of the enterprise, and are difficult to quantify, though there is a large body of literature exhorting businesses to adopt measures beyond financial ones.

8.6.1 Human Resources: A CSR programme can be an aid to recruitment and retention, particularly within the competitive graduate student market. Potential recruits often ask about a firm's CSR policy during an interview, and having a comprehensive policy can give an advantage. CSR can also help improve the perception of a company among its staff, particularly when staff can become involved through payroll giving, fundraising activities or community volunteering. CSR has been found to encourage customer orientation among frontline employees.

8.6.2 Risk Management: Managing risk is a central part of many corporate strategies. Reputations that take decades to build up can be ruined in hours through incidents such as corruption scandals or environmental accidents. These can also draw unwanted attention from regulators, courts, governments and media. Building a genuine culture of *doing the right thing* within a corporation can offset these risks.

The desire of enterprises to improve their risk management is a powerful factor behind CSR. Enterprises generally agree that CSR helps them in managing their risks, their intangible assets, their internal processes, and their relations with internal and external stakeholders. Although most businesses support the assumption of a positive impact of CSR on competitiveness, particularly in the long-term, they are however not able to quantify this effect.

8.6.3 Brand Differentiation: In crowded marketplaces, companies strive for a unique selling proposition that can separate them from the competition in the minds of consumers. CSR can play a role in building customer loyalty based on distinctive ethical values.

8.6.4 Relationship between Enterprises and Society: The potential of CSR policies to strengthen the symbiotic relationship between enterprises and society needs to be tapped in areas such as sustainable growth, education and social cohesion. CSR can support the creation of an atmosphere of trust within companies, which leads to a stronger commitment of employees and higher innovation performance. A similar atmosphere of trust in co-operation among other stakeholders (business partners, suppliers, and consumers) can increase the external innovation performance. Consumer confidence fostered through CSR can be a major contributor to economic growth. More specifically, through CSR practices, enterprises can play an important role in preventing and combating corruption and bribery, and discouraging the use of enterprises for money laundering and criminal activities financing.

CSR policies can also boost the societal benefit that enterprises create with regard to innovation. Innovative practices aiming at better jobs, safer and employee-friendly workplaces, gender mainstreaming and the innovation or technology transfer to local communities and developing countries, leading to a more equitable economic and social development, are further examples of societal benefits created by innovative enterprises. Indeed, CSR can play a positive role in fostering development by helping to establish a dialogue among public authorities, social partners, civil society and companies.

8.6.5 Business-to-business Good Practices: The exchange of experience and good practices about CSR between companies, can serve as an important tool to develop the concept further. It can help businesses to acquaint themselves with the concept, to benchmark their position against competitors and to build up a consensus about its instruments, such as reporting standards or verification procedures. These exchanges could be particularly beneficial at sectoral level, where they can play an important role is identifying common challenges and options for co-operation among competitors. Such co-operation could reduce the costs of adopting CSR and help to create a level-playing field.

8.6.6 Developing Management Skills: The importance of education and training of managers, employees, and other actors to promote CSR cannot be undermined. The education system, at all levels, has a crucial role to play in the fostering of social responsibility in citizens, including those who are working—or will work—in the world of business or outside of it. It can fulfil this role by enabling citizens to understand and appreciate social, environmental and ethical values and equipping them to take informed decisions. Education and training in the field of business administration have particular relevance to CSR in this context, and the encouragement of an effective dialogue between the worlds of business and education on this subject can contribute to the promotion of CSR principles and practices.

8.6.7 Cleaner Technologies: One of the fundamental objectives under CSR includes sound and progressive environmental principles. Experience shows that undertakings, which follow sound environmental standards are, in medium- and long-term, more competitive in the international market, as environmental standards promote innovation and modernisation of processes and products and lead to cleaner technologies. Conversely, it may also be that internationally competitive firms are better able to invest in newer, more efficient and cleaner technologies.

The importance of environmental technologies, and their ability to contribute both to the profitability of individual firms and to the competitiveness of the economy has been highly stressed. In particular, the use of cleaner 'integrated' technologies are often associated with improved process efficiency and can provide savings for the firm, thanks to the improved resource efficiency, reduced waste etc. This provides clear guidance for action by enterprises willing to improve their environmental performance.

CSR has partly evolved in response to consumer demands and expectations. Consumers, in their purchasing behaviour, increasingly require information and reassurance that their wider interests, such as environmental and social concerns, are being taken into account. Enterprises are increasingly sensitive to these demands both to retain existing customers and to attract new customers.

Consumers and their representative organisations have therefore an important role to play in the evolution of CSR. If CSR is to continue to serve its purpose, strong lines of communication between enterprises and consumers need to be created.

8.6.8 Other Benefits: In addition to the above-mentioned, an organization can reap the following benefits out of CSR:

1. It can enhance brand image and reputation.
2. There have been enough instances where CSR has added new customers to companies. Typical example would be health awareness programmes which generate need for related products.
3. CSR not only adds to the reputation of the brand but also gives confidence to the customers that they are dealing with responsible, long-term institutions and not short-term operators.

4. CSR can forge partnerships with other organizations.

Thus, an increasing number of firms have embraced a culture of CSR. Only competitive and profitable enterprises are able to make a long-term contribution to sustainable development by generating wealth and jobs without compromising the social and environmental needs of the society. In fact, only profitable firms are sustainable and have better chances to adopt/develop responsible practices.

The role of enterprise policy is to help create a business environment, which supports sustainable economic growth. Its objective is to ensure a balanced approach to sustainable development, which maximises synergies between its economic, social and environmental dimensions. Another key element is to support businesses in enhancing their competitiveness and in meeting the challenges of the transition to the knowledge economy.

8.7 Convergence and Transparency of CSR Practices

The need for transparent and proactive communication of CSR is a key issue of concern. Much of the demand for CSR disclosure may be viewed as a result of public desire for information on which to base an opinion about whether or not a corporation is appropriate or right and proper, i.e. to evaluate corporate legitimacy. Many corporations voluntarily disclose social information. These disclosures may take the form of management discussion in annual reports, or separate disclosures in social, sustainability or environmental reports. Largely, the format, content and detail of such disclosures are unregulated.

CSR relates to a very wide range of company activities. This is particularly the case when an enterprise operates in several countries and has to adapt its activities to the specific situations in these countries. This diversity has helped to create an impressive richness of voluntary enterprise initiatives, which often include innovative elements, but also implies challenges, namely the lack of transparency and comparability.

Transparency is a key element of the CSR debate as it helps businesses to improve their practices and behaviour. Transparency also enables businesses and third parties to measure the results achieved. CSR benchmarks against which the social and environmental performance of businesses can be measured and compared are useful to provide transparency and facilitate an effective and credible benchmarking. The interest in benchmarks has resulted in an increase in guidelines, principles and codes in recent years.

Several market-driven international multi-stakeholder initiatives are emerging, which work towards convergence and transparency in the area of CSR. The areas where initiatives have been taken to promote CSR are as under:

1. Codes of conduct.
2. Management standards.
3. Socially responsible investment.

8.7.1 Codes of Conduct: The increasing public interest in social and environmental impact and ethical standards of industry has moved many companies, in particular those of the consumer goods sector, to adopt codes of conduct relating to labour issues, human rights and the environment.

Codes of conduct are innovative and important instruments for the promotion of fundamental human, labour and environmental rights, and anti-corruption practices, especially in countries where public authorities fail to enforce minimum standards.

The biggest challenge related to codes is to ensure that they are effectively implemented, monitored and verified. Special attention should be given to implementing codes in respect of workers in the informal sector and sub-contractors and in free-trade zones.

Codes of conduct should include appropriate mechanisms for evaluation and verification of their implementation, as well as a system of compliance.

8.7.2 Management Standards: Faced with a widening range of complex issues in areas such as labour practices and supplier relations, businesses, regardless of sector, size, structure or maturity, would benefit from the inclusion of social and environmental issues into their daily operations. In this context, CSR management systems—like total quality management systems—could allow enterprises to have a clear picture of their social and environmental impacts, and help them to target the significant ones and manage them well.

The Eco-management and Audit Scheme (EMAS), for example, allows voluntary participation in an environmental management scheme. It is a scheme for companies and other organisations that are willing to commit themselves to evaluate, manage and improve their environmental and economic performance. In addition, active involvement of employees is a driving force for EMAS and a contribution to the social management of organisations.

8.7.3 Socially Responsible Investment (SRI): For SRI to contribute to the promotion of CSR, the development by rating organisations of criteria and indicators which identify the factors of competitive advantage and business success of socially responsible enterprises is essential.

Registration statements and prospectuses issued at the time of an IPO (initial public offering), can be another useful source of information on social and environmental risks allowing the prospective investor to assess the overall risks associated with a business.

Deeply-rooted societal changes such as increasing participation of women in the labour market should be reflected in CSR, adapting structural changes and changing the work environment in order to create more balanced conditions for both genders acknowledging the valuable contribution of women as strategies which will benefit the society as well as the enterprise itself.

SRI has come to stay with the creation of many funds that invest only in companies that exhibit socially responsible behaviour and preliminary evidence suggests that such funds are providing excellent returns to their customers.

Recently, institutions like the Global Reporting Initiative have come out with elaborate guidelines for preparing social or sustainability reports. These guidelines have been taken up by many organizations as a framework to build their social reports. In Asia, Japan has made significant strides in building such corporate social reporting initiatives, but other countries like India and China have a very limited number of companies, and that too in select industries like oil, chemicals and steel, providing social reporting. However, the contents of such reporting, particularly in countries like India, remain relatively understudied. Many aspects of CSR reporting in these countries still remain unintelligible. In most of the studies, the areas of social responsibility have been identified as environment, equal opportunity, community involvement and development and each one of these areas has been further sub-divided into specific indicators.

CSR-oriented financial institutions have robust stakeholder engagement programmes to solve complex sustainability challenges, provide inputs into business strategy, and keep abreast of stakeholder concerns. While the debate starts to gather steam over the degree to which financial institutions have a responsibility to advance sustainability, best practice CSR financial institutions are playing a lead role in promoting sustainable practices within their operations, and to their consumers and the public at large. Best practice CSR financial institutions are treating sustainability as a business strategy and opportunity—not as an add-on, feel-good charitable endeavour.

They are developing strategic CSR policies and programmes across a range of social and environmental areas to reduce or offset their negative, and boost their positive, direct and indirect impacts.

The leading financial institutions (FIs) are implementing following measures to promote CSR:

- Setting up governance and management systems to embed CSR into their operations.
- Reporting on their CSR performance with independent validation.

- Developing a CSR product array that integrates social and environmental considerations.
- Integrating CSR into their risk management systems.
- Developing methods for screening their investments against CSR criteria.

8.8 Globalisation and CSR

Globalisation implies widening and deepening integration with the *globe,* i.e. with people and processes abroad. The trend towards the evolution of a global society is generally thought of in economic terms and in terms of the consequences of the revolution in communication technologies. There is undoubtedly much greater economic integration among the nations of the world today. Globalisation is widely seen as the most important factor that could influence economies of nations the world over in the new millennium. The rapid advancement in information technology and communications has made it not just possible but absolutely essential for economies of the world to adapt or fall by the wayside.

8.8.1 Dimensions and Features of Globalisation: The 1990s witnessed a paradigm shift towards market-oriented economic policies and a careful dismantling of obstacles in its wake. This has helped the smaller, emerging economies gain access to world markets, emergent technologies and collaborations. This has also given them a window to the developed world and helped them understand the significant role of globalisation as an instrument, which could be utilised not just to achieve economic efficiency, but also eradicate poverty.

Globalisation has also resulted in the creation of a new business framework. More changes can be expected in the business scenario specifically in terms of openness, adaptiveness and responsiveness. The most important dimensions of economic globalisation are: (a) breaking down of national barriers; (b) international spread of trade, financial and production activities and (c) growing power of transnational corporations and international financial institutions in these processes. While economic globalisation is a very uneven process, with increased trade and investment being focused in a few countries, almost all countries are greatly affected by this process.

A major feature of globalisation is the growing concentration and monopolisation of economic resources and power by transnational corporations and by global financial firms and funds. This process has been termed transnationalisation, in which fewer and fewer transnational corporations are gaining a large and rapidly increasing proportion of world economic resources, production and market shares. Where a multinational company used to dominate the market of a single product, a big transnational company (TNC) now typically produces or trades in an increasing multitude of products, services and sectors. Through mergers and acquisitions, fewer and fewer of these TNCs now control a larger and larger share of the global market, whether in commodities, manufactures or services.

Another feature of the current globalisation process is the globalisation of national policies and policy-making mechanism. National policies (including economic, social, cultural and technological) that until recently were determined by the States and people within a country have increasingly come under the influence of international agencies and processes or by big private corporations and economic/financial players. This has led to the narrowed ability of governments and people to make choices from options in economic, social and cultural policies.

Most developing countries have seen their independent policy-making mechanism capacity eroded, and have to adopt policies influenced by other entities, which may on balance be detrimental to the countries concerned. The developed countries, where the major economic players reside, and which also control the processes and policies of international economic agencies, are better able to maintain control over their own national policies as well as determine the policies and practices of international institutions and the global system. However, it is also true that the large corporations have taken over a large part of decision-making even in the developed countries, at the expense of the power of the state or political and social leaders.

8.8.2 Global Dimension of CSR: Global governance and the interrelation between trade, investment and sustainable development are key issues in the CSR debate. Indeed, awareness of CSR issues and concerns will contribute to promote more sustainable investments, more effective development co-operation and technology transfers. In the new global economy, companies that are responsive to the demands of all of their stakeholders are arguably better positioned to achieve long-term financial success. It is no longer optional for a company to communicate its environmental and social impacts; such information is pertinent in an information-driven economy, and improved communication has become critical for sustainable business growth.

Both processes of trade and financial markets liberalisation should be matched by appropriate progress towards an effective system of global governance including its social and environmental dimensions. Globalisation has also increasingly exposed enterprises to trans-boundary economic criminality, requiring an international response.

By abiding by internationally accepted standards, multinational enterprises can contribute to ensure that international trade markets function in a more sustainable way and it is therefore important that the promotion of CSR at international level takes as its basis international standards and agreed instruments. At the same time, identifying common frameworks for the global dimension of CSR is challenging due to the diversity in domestic policy frameworks, protection of workers and environmental regulation.

8.9 Promoting CSR: The Challenges

The challenges to CSR practices among enterprises stem

from insufficient knowledge about the relationship between CSR and business performance. Also, there is lack of consensus among various parties involved on an adequate concept taking account of the global dimension of CSR, in particular the diversity in domestic policy frameworks in the world. Besides, there is lack of teaching and training about the role of CSR, especially in commercial and management schools.

Focussing the strategy on the following areas would help in wider diffusion of CSR:

- Increasing knowledge about the positive impact of CSR on business and societies.
- Developing the exchange of experience and good practices on CSR between enterprises.
- Promoting the development of CSR management skills.
- Integrating CSR into community policies.

CSR has found recognition among enterprises, policy-makers and other stakeholders, as an important element of new and emerging forms of governance, which can help them to respond to the following fundamental changes:

1. Globalisation has created new opportunities for enterprises, but it has also increased their organisational complexity and the increasing extension of business activities abroad has led to new responsibilities on a global scale, particularly in developing countries.
2. Considerations of image and reputation play an increasingly important role in the business competitive environment, as consumers and NGOs ask for more information about the conditions in which products and services are generated and the sustainability impact thereof, and tend to reward, with their behaviour, socially and environmentally responsible firms.
3. Partly as a consequence of this, financial stakeholders ask for the disclosure of information going beyond traditional financial reporting so as to allow them to better identify the success and risk factors inherent in a company and its

responsiveness to public opinion.

4. As knowledge and innovation become increasingly important for competitiveness, enterprises have a higher interest in retaining highly skilled and competent personnel.

8.10 Criticisms and Concerns

Critics of CSR as well as proponents debate a number of concerns related to it. These include relationship of CSR to the fundamental purpose and nature of business and questionable motives for engaging in CSR, including concerns about insincerity and hypocrisy.

Proponents argue that corporations make more long-term profits by operating with a perspective, while critics argue that CSR distracts from the economic role of businesses. Some argue that CSR is merely window-dressing, or an attempt to pre-empt the role of governments as a watchdog over powerful multinational corporations.

It is argued that a corporation's purpose is to maximize returns to its shareholders. Corporations are only responsible to their shareholders and not to society as a whole. Some people perceive CSR as incongruent with the very nature and purpose of business, and indeed a hindrance to free trade.

Critics of this argument perceive free market as opposed to the well-being of society and a hindrance to human freedom. They claim that the type of capitalism practiced in many developing countries is a form of economic and cultural imperialism, noting that these countries usually have fewer labour protections, and thus their citizens are at a higher risk of exploitation by multinational corporations.

Critics concerned with corporate hypocrisy and insincerity generally suggest that better governmental and international regulation and enforcement, rather than voluntary measures, are necessary to ensure that companies behave in a socially responsible manner.

9

Corporate Social Responsibility (CSR) in India

A movement is growing around the world, within businesses themselves, to develop frameworks of accountability that go beyond the responsibility of business towards investors and customers to responsibility towards citizens and society. Business organisations are expected to report for public scrutiny, the impact of their business activities on the environment and communities. Leading Indian companies are part of this international movement.

This voluntary movement for accountability must be strengthened. It must spread beyond a few leading firms. Business organizations do not want more government control of their activities. They resent government imposed reporting requirements, especially those that are tedious and costly to implement. However, reporting standards developed by business associations themselves, as the new frameworks are, should be acceptable. Government could work with industry associations and put pressure for the wider adoption of such standards. They should not be footnotes to financial accounts. They must be front and centre where they are noticed.

Private business enterprises are the principal engines of industrialisation in India. Government is no longer promoting public sector enterprises in industry. Its principal role in industry today is to enable and regulate the activity of business enterprises in the private sector to ensure that they may profit and grow, and meet societal needs too. Therefore, Government must induce private enterprises to pay more attention to societal concerns of inclusion and environmental sustainability.

9.1 Pressure for CSR

Now-a-days across the globe, companies are under

massive pressure from their shareholders, investors, media, NGOs as well as their customers to carry out business in a responsible and ethical manner. Recently, several initiatives have been initiated to promote CSR practices around the globe. For instance, United Nations Environment Programme-Finance Initiative (UNEP-FI) was launched in 1990s to promote sustainable development within the framework of market mechanisms. Likewise, Bank Tract Network was formed by global coalition of NGOs to promote sustainable finance. State regulatory bodies, media, NGOs, and customers have significantly addressed social responsibility issues in the banking sector.

Likewise, international organizations such as the World Bank also frequently exert pressures on banks to analyse social and environmental risks involved in projects to be financed. In addition to this, CSR practices of an organization have significant impact on its reputation and resultant profitability. Many researchers have found positive correlation between CSR and the financial performance of an organization. Some researchers have also proved a negative impact of socially irresponsible operation on share prices and brand reputation of a bank. Around the globe, banking industry is showing a good commitment to CSR principles. Banks have exhibited conscious efforts to comply with the relevant provisions to reduce the regulatory action through depicting a good environmental citizen's image.

9.1.1 Public Policies: CSR has inspired national governments to include CSR issues into their national public policy agendas. The increased importance driven by CSR has prompted governments to promote socially and environmentally responsible corporate practices. Over the years, governments have considered CSR as a public issue that requires national governmental involvement to address the very issues relevant to CSR. The heightened role of government in CSR has facilitated the development of numerous CSR programmes and policies.

While the interests of shareholders and the actions of

managers of any business enterprise have to be governed by the laws of economics, requiring an adequate financial return on investments made, in reality, the operations of an enterprise need to be driven by a much larger set of objectives that are today being defined under the term CSR.

The broad rationale for a new set of ethics for corporate decision-making, which clearly constructs and upholds a company's social responsibility, arises from the fact that a business enterprise derives several benefits from the society, which must, therefore, require the enterprise to provide returns to society as well.

Of course, the system of taxation in most countries does ensure that basic services provided by government—such as a system of law and order, provision of infrastructure that includes assets such as roads, transportation facilities, the benefits received from the apparatus of society for respecting and enforcing property rights etc.—are paid for through taxation on economic goods and services produced and consumed. However, there are other aspects of services provided by the society that have now become even more important than traditional relationship between government and businesses. These go far beyond what was the case a few decades ago.

9.2 CSR and Disaster Preparedness

The concept of corporate social responsibilities is generally understood to mean that corporations have a degree of responsibility not only for the economic consequences of their activities but also their social and environmental implications. This is sometimes referred to as a triple bottom line approach that considers the economic, social and environmental aspects of corporate activities. However, the corporates have, apart from these obligations, to work for the people on humanitarian aspects also such as relief, rescue, and recovery, reconstruction during various anthropogenic and natural disasters etc.

Disaster is an event, be it natural or man-made that

threatens the whole society. In regards, risk reduction and mitigation of disasters is very important because human lives, properties and human rights as a whole are lost when it strikes. Management of disasters in general and risk reduction of disasters in particular imply that it is not a quick fix mechanism, it needs meticulous planning and coordination among various role players like government, non-governmental agencies, community and corporate bodies. Among all disasters that occur in Bihar, floods of Koshi River are the most frequent and often, most devastating ones. The factors of floods in Bihar are mainly due to the peculiarity of rainfall and improper structural management practices of relevant responsible stakeholders.

9.3 Relevance of CSR in India

The main function of an enterprise is to create value by producing goods and services that society demands, thereby generating profit for its owners and shareholders as well as welfare for the society, particularly through an ongoing process of jobs creation. However, new social and market pressures are gradually leading to a change in the values and in the horizon of business activity.

CSR is a concept whereby companies integrate social and environmental concerns in their business operations and in their interaction with their stakeholders on a voluntary basis. CSR is the point of convergence of various initiatives aimed at ensuring socio-economic development of the community which would be livelihood-oriented as a whole in a credible and sustainable manner.

CSR developed very slowly in India though it was started a long time ago. CSR has been assuming greater importance in the corporate world, including the banking sector. There is a visible trend in the financial sector for promoting environment-friendly and socially responsible lending and investment practices.

Presently, there is a growing perception among enterprises

that sustainable business success and shareholder value cannot be achieved solely by maximising short-term profits, but instead through market-oriented, yet responsible behaviour. Companies are aware that they can contribute to sustainable development by managing the operations in such a way as to enhance economic growth and increase competitiveness while ensuring environmental protection and promoting social responsibility, including consumer interests.

The increasing relevance of CSR in India has stemmed from the fact that a business cannot succeed by ignoring the human and social needs of a society. In this age of widespread communication and growing emphasis on transparency, customers of any product or service are unlikely to feel satisfied in buying from a company that is seen to violate the expectations of ethical and socially responsible behaviour.

Therefore, companies that pay genuine attention to the principles of socially responsible behaviour are favoured by the public and preferred for their goods and services.

In the Indian context, protection of the environment and country's natural resources would certainly be of paramount importance in understanding this concept with regards to sustainable development. Nevertheless, what would be equally important is the need to ensure that society does not suffer from disparities of income and provision of basic services like health care, education and literacy.

Consequently, if corporate actions are to target the most fundamental problems (poverty, illiteracy etc.), they become guideposts for corporate social strategy and action.

It is important to mention here that the progress and welfare of society is not merely the responsibility of governments alone. In an effective sense, it involves appropriate actions by all stakeholders, of which the corporate sector is extremely important. Hence, actions to address some of these basic challenges also become important for leaders of business and industry.

The objectives of sustainable development rest within the

principles of CSR, because unless the needs of society, both present and future, are served, sustainable development would remain only a myth. And the most significant step in pursuing CSR is to proactively protect the environment.

CSR in India is still characterized mainly by philanthropic and community development activities and Indian companies and stakeholders have begun to adopt some aspects of the mainstream agenda, such as the integration of CSR into their business processes and engagement in multi-stakeholder dialogues.

A lot of studies on CSR practices and perceptions have been undertaken in the context of Indian economy and its different sectors have attempted to map the corporate social responsibility practices of major players in the Indian banking and found that CSR has a positive impact on the performance and image of banks.

In a globalised economy, the industries and firms are vulnerable to stringent environmental policies, severe law suits or consumer boycotts. Since banking sector is one of the major stake holders in the industrial sector, it can find itself faced with credit risk and liability risks.

Further, environmental impact might affect the quality of assets and also the rate of return of banks in the long-run. Thus, the banks should go green and play a pro-active role to take environmental and ecological aspects as part of their lending principle, which would force industries to go for mandated investment for environmental management, use of appropriate technologies and management systems.

9.4 Initiatives of Reserve Bank of India (RBI)

RBI has suggested that banks should pay special attention towards integration of social and environmental concerns in their business operations to achieve sustainable development. RBI has also pointed out to the banks to start non-financial reporting (NFR) which should cover the work done by the banks towards the social, economic and environmental betterment of the society. CSR in Indian banking sector is aimed towards addressing financial inclusion, providing financial services to the unbanked or untapped areas of the country, and the socio-economic development of the country by focusing on activities like poverty eradication, health and medical care, rural area development, self-employment training and financial literacy trainings, infrastructure development, education, and environmental protection etc. Against this backdrop, it can be understood that CSR is now an integral part of functioning of an organization. The performance of an organization should be judged beyond the financial parameters.

However, it is important to mention that in spite of mass acceptance to the concept of CSR reporting by banking sector around the globe, Indian banks are not responding in the expected manner. According to some critics, Indian financial institutions are concerned, to the exclusion of all other considerations, about the ecology of their balance sheets and, therefore, focused on ever-greening their assets.

To highlight the role of banks in CSR, RBI circulated a notice on December 20, 2007 for all the scheduled commercial banks in India. Major issues discussed in the notice were regarding CSR, sustainable development (SD) and non-financial reporting (NFR). Briefing about CSR programme to member commercial banks, RBI followed many international initiatives to highlight the importance of this notice like United Nations Environment Programme-Finance Initiative (UNEP-FI), Global Reporting Initiative (GRI), Equator Principles, and Declaration on Financial Institutions. Apart from these international initiatives, RBI report also talked about other important and urgent issues regarding global warming and extent of the problem, the economics of climate change, Happy Planet Index, Kyoto Protocol etc. and advised the banks to implement the same earnestly and sincerely.

In the context of Indian financial sector, very little systematic documentation of CSR initiatives is available so far. It is observed that Indian financial institutions are found to be adopting an integrated approach of combining CSR with the ultimate customer satisfaction voluntarily.

The major thrust areas for CSR practice in Indian banks are common in public sector and private sector banks. These areas include children welfare, community welfare, education, environment, healthcare, poverty eradication, rural development, vocational training, women's empowerment, protection to girl child and employment.

The contribution of financial institutions including banks to sustainable development is paramount, considering the crucial role they play in financing the economic and developmental activities of the world. In this context, the urgency for banks to act as responsible corporate citizens in the society, especially in a developing country like India need to be hardly over emphasized. Their activities should reflect their concern for human rights and environment. Global warming and climate change are particularly important in the context of sustainable development, especially for developing countries, which tend to be ill-equipped for such changes. Therefore, all concerned should come forward to check the situation.

Financial institutions in India are showing interest in integrating sustainability into their business models but its CSR reporting practices are far from satisfaction. There are only few banks which report their activities on triple bottom line principles. As a matter of fact, the standards for rating CSR practices are less uniform in comparison to that for financial rating. Absence of mandatory provisions regarding reporting of CSR practices further cause negligence on the issue of reporting. The adaptation of reporting CSR practices, though quite slow in India, would definitely get a great pace in near future. In fact, there is a great need for enacting some stringent regulatory provisions to ensure the adherence to social responsibility principles.

The recent initiatives of RBI in this regard are expected to bring a drastic change in the near future. Further, voluntary actions are required to be taken by the financial bodies to ensure the socio-environmental feasibility of projects to be financed. Indian banking sector must also depict its socially responsible behaviour through integrating triple bottom line principle. Banks must also provide appropriate training to their employees on environmental and social risks to ensure that climate change is taken into account in corporate banking decisions. Further, regulatory authorities must envisage the regulations for initiating reforms in reporting practices.

9.5 CSR in Companies Act, 2013

The 2013 Act makes an effort to introduce the culture of corporate social responsibility (CSR) in Indian corporates by requiring companies to formulate a corporate social responsibility policy and at least incur a given minimum expenditure on social activities.

It may be recalled that the Ministry of Corporate Affairs (MCA) had introduced the Corporate Social Responsibility Voluntary Guidelines in 2009. These guidelines have now been incorporated within the 2013 Act and have obtained legal sanctity. Section 135 of the 2013 Act, seeks to provide that every company having a net worth of ₹ 500 crore, or more or a turnover of ₹ 1,000 crore or more, or a net profit of ₹ 5 crore or more, during any financial year shall constitute the corporate social responsibility committee of the board. This committee needs to comprise of three or more directors, out of which, at least one director should be an independent director. The composition of the committee shall be included in the board's report. The committee shall formulate the policy, including activities specified in Schedule VII, which are as follows:

1. Eradicating extreme hunger and poverty.
2. Promotion of education.
3. Promoting gender equality and empowering women.
4. Reducing child mortality and improving maternal health.
5. Combating human immunodeficiency virus, acquired

immune deficiency syndrome, malaria and other diseases.

6. Ensuring environmental sustainability.
7. Employment enhancing vocational skills.
8. Social business projects.
9. Contribution to the Prime Minister's National Relief Fund or any other fund set up by the central government or the state governments for socio-economic development and relief, and funds for the welfare of the scheduled castes and tribes, other backward classes, minorities and women.
10. Such other matters as may be prescribed.

There have been mixed reactions to the introduction of the 'spend or explain' approach taken by the MCA with respect to CSR. It may take a while before all of Corporate India imbibes CSR as a culture.

However, activities specified in the Schedule are not elaborate or detailed enough to indicate the kind of projects that could be undertaken, for example, environment sustainability or social business projects could encompass a wide range of activities.

The committee will also need to recommend the amount of expenditure to be incurred and monitor the policy from time-to-time. The board shall disclose the contents of the policy in its report, and place it on the website, if any, of the company. The 2013 Act mandates that these companies would be required to spend at least 2 percent of the average net profits of the immediately preceding three years on CSR activities, and if not spent, explanation for the reasons thereof would need to be given in the director's report (Section 135 of the 2013 Act).

In short, Companies Act, 2013 is a landmark legislation with far reaching consequences on all companies incorporated in India. Corporate India continues to evolve at a fast pace leading to the emergence of a diverse set of stakeholders. The growth has led to a surge in consequences such as risk and default that seem to be visibly impacting the virtues of governance, which is precisely the area on which the Act promises to substantively raise the bar. The Act is also quite outward looking and in several areas attempts to harmonize with international requirements. The Act in a comprehensive form purports to deal with various aspects of corporate India and Indian companies will have to closely examine these developments to develop a clear strategy at ensuring compliance as per the new requirements.

9.6 Some Concerns

Many factors, within general business practices, are altering to ensure that every person benefits from the continued functioning of the company. Previously many businesses have subscribed to practices that may have had negative effects on their stakeholders. This is now changing as the realization has set in, of the true importance of the different stakeholders in any particular business. There are many different manners in which a company can implement corporate social responsibility measures for the benefit of all concerned. The manner in which each different company implements the changes will be dependent on what aspects of the company could be considered as having produced negative effects.

Corporate social responsibility can often be confused with corporate charity, but it is a very different thing. Corporate charity can involve the donation of money and the provision of opportunities to members of the community and stakeholders. This is very different to the considerations that a company must abide by, to ensure that their actions fall within the acceptable corporate social responsibility guidelines that have been established. These can include guidelines that relate to the environmental impact that a particular action can have or they can relate to the impact that an action can have on the local community. The guidelines are intended to ensure that any negative effects that an action could possibly have are eliminated or reduced as far as possible.

The concept of corporate social responsibility is intended to provide each business with a far greater ability to create sustainable development. This can help to provide ongoing benefits to the business and its stakeholders, regardless of the size of the business. The business can become more competitive once it begins to create and follow guidelines to enforce the concept of corporate social responsibility within the company and its surrounding community. The greater the sustainability of a business's developments, the more successful it can become. The attraction of corporate social responsibility for many businesses is that it can help to increase sustainability without creating negative effects.

The type of business that implements changes to come in line with accepted corporate social responsibility guidelines will also determine what sort of changes will need to be made. The size of the investment that will be required to ensure that the business conforms to acceptable guidelines will also be affected by the type of business it is. The resources that are expended for this purpose do need to be viewed as an investment rather than an unnecessary expense since the positive results can more than repay the investment.

The term corporate social responsibility may seem to be one of the current buzz words in business, but the concept is one that has been formulated by many people over many years. The concept is one that is seen as being important for the continued well-being of the vast majority of companies and essential for the community that surrounds them. The importance that is placed on business actions conforming to acceptable corporate social responsibility standards can ensure that these guidelines are treated with the respect that they merit.

Supported and nurtured leadership among self-help groups and village advocacy groups, encouraged multi-cropping and irrigation across thousands of acres in over 700 villages, earmarked dedicated resources for social welfare, and rural and disaster management programmes in Bihar.

The concept of corporate social responsibilities is generally understood to mean that corporations have a degree of responsibility not only for the economic consequences of their activities but also their social and environmental implications. This is sometimes referred to as a triple bottom line approach that considers the economic, social and environmental aspects of corporate activities. However, the corporate has, apart from these obligations, to work for the people on humanitarian aspects such as relief, rescue, and recovery, and reconstruction during various anthropogenic and natural disasters.

Disaster is an event, be it natural or man-made that threatens the whole society. In regards, risk reduction and mitigation of disasters is very important because the human lives, properties and human rights as a whole are lost when it strikes. Management of disasters in general and risk reduction of disasters in particular imply that it is not a quick fix mechanism, it needs meticulous planning and coordination among various role players like government, non-governmental agencies, community and corporate bodies.

To sum up, while in the mid-1990s, local authorities were probably the most active players trying to achieve sustainable development, the focus has recently shifted to business as a major actor. Many responsible business managers and their firms have opted for eco-efficiency as their guiding principle. Eco-efficiency is the economic value added by a firm in relation to its aggregated ecological impact. The World Business Council for Sustainable Development (WBCSD) has defined eco-efficiency as follows: 'Eco-efficiency is achieved by the delivery of competitively priced goods and services that satisfy human needs and bring quality of life, while progressively reducing ecological impacts and resource intensity throughout the life-cycle, to a level at least in line with the earth's carrying capacity'.

Similar to the concept of eco-efficiency, but so far less explored in corporate sustainability is the concept of socio-efficiency, that is, the relation between a firm's value added and its social impact. While it can be assumed that corporate impact on environment is usually negative, this may not be true for the social impact. Depending on the type of socio-efficiency, one can either try to minimize the negative social impact, or maximize the positive social impact while pursuing the value-added activity. Both eco-efficiency and socio-efficiency promote economic sustainability of the businesses in the long-run.

As a result of increased consciousnesses towards sustainability issues, a move from core financial reporting to the adoption of *triple bottom line principle* has been observed in recent years. The triple bottom line principle entails reporting of not only the economic performance of company but also their environmental and social performance, i.e. reporting of CSR activities. CSR reporting calls for reflection of corporate ethical practices, transparency, sensitivity to the environmental issues, social commitment and labour welfare practices of business houses.

10

Disaster Management in Select Countries

10.1 USA

In the USA, the Federal Emergency Management Agency (FEMA), Department of Homeland Security, is the nodal agency for disaster management. FEMA administers various programmes for disaster mitigation response and recovery under the public assistance (PA) grant programme. Assistance is provided to states and tribal local bodies to enable communities to quickly respond to and recover from major disasters or emergency declared by the President. Under this programme, assistance is provided for debris removal, emergency protection measures, as well as repair, replacement, or restoration of disaster damaged, publicly-owned facilities. The federal share of assistance is not less than 75 percent of the eligible cost for emergency majors and permanent restoration. FEMA also implements many disaster-specific assistance programmes like fire management assistance grant programme, flood mitigation assistance programme, national earthquake hazards reduction programme and repetitive flood claims programme.

FEMA also implements mitigation programmes like the Hazard Mitigation Grant Programme (HMGP) that provides grants to states and local governments to implement long-term hazard mitigation measures after declaration of a major disaster. The purpose of the HMGP is to reduce loss of life and property due to natural disasters and to enable mitigation measures to be implemented during immediate recovery from the disaster.

10.2 Canada

Public Safety (PS) Canada is the nodal agency for disaster management in Canada. In the event of a large natural disaster, the Government of Canada provides financial assistance to provincial and territorial governments through the Disaster Financial Assistance Arrangements (DFAA), which is administered by PS Canada. The Government of Canada bears up to 90 percent of the relief expenditure on a graded basis. Under this arrangement, the government supports evacuation operations, restoration of public works and infrastructure, replacement or repair of basic essential personal property of individuals etc. The Emergency Act of Canada has been put in place to strengthen the emergency management activities in the country. It sets out clear rules and responses for all federal ministries across the full spectrum of emergency management.

10.3 Australia

In Australia, Emergency Management Australia (EMA) is the nodal agency for disaster management at the federal level. Natural disaster management is constitutionally a responsibility of the state or territory and EMA offers various programmes for effectively mitigating, responding to, and recovering from their natural disasters. The Australian Government provides funding through the Natural Disaster Relief and Recovery Arrangement (NDRRA), which is administered by EMA on its behalf. Under the arrangement, a state or territory may claim NDRRA funding, if it has spent more than a specified amount on relief and recovery expenditure in case of a natural disaster. The amount of NDRRA funding would depend on a pre-defined threshold derived on the state's revenue. The NDRRA applies to natural disasters like flood, storm, earthquake, cyclone, landslide, tsunami and the like, but does not apply to 'other unspecified events like drought, frost, heat wave, epidemic' etc.

The Australian Government also has a Natural Disaster Mitigation Programme which is aimed at identifying and addressing natural disasters in order of risk priority across the nation. Under this scheme, the Federal Government generally contributes up to one-third of the costs of the project, other than certain specific projects like installation of flood warning systems, infrastructure upgrades, etc., where it bears half the cost. 50 percent of the central share is paid in advance by the Australian Government and the remaining 50 percent is paid out in quarterly instalments on receipt of claims from the states.

10.4 South Africa

The Disaster Management Act, 2002 of South Africa provides for national, provincial, and municipal disaster management centres. The primary responsibility of disaster management lies with the local and the provincial governments. However, depending upon the intensity of the disaster, the National Government may intervene and provide adequate financial assistance. The Act provides for declaration of a disaster as local, provincial or national disaster. For the purpose of immediate relief, it maintains a Disaster and Emergency Fund, which is used to supplement the efforts of the local and provincial governments and is operated by the central cabinet.

Activities relating to post-disaster reconstruction are funded from a national reserve established in line with budgetary requirements under the medium-term expenditure framework to provide contingency funds for a range of situations. The Act deals only with preparedness, response, and recovery and leaves mitigation to be taken up by the respective ministries from their budgeted grants.

10.5 Japan

In Japan, the basic framework of disaster management is provided by the Disasters Countermeasures Basic Act, 1961. The government provides various grants and loans to the prefectural and municipal governments for their effort in meeting response and recovery expenditure.

10.6 Brazil

In Brazil, the nodal agency for disaster management is the National Civil Defence Secretariat. The Secretariat works in coordination with the state and municipal governments in case of activities relating to civil defence. These activities are funded, at the central level, out of the Civil Defence Action Programme of the Union Budget. The respective states and municipalities have similar budgetary provisions to meet the expenditure on disaster management.

Appendix 1

Classification of Disasters

Disasters are not new to mankind. They have been the constant, though inconvenient, companions of the human beings since times immemorial. Disasters can be natural or man-made. Earthquakes, cyclones, hailstorms, cloud-bursts, landslides, soil erosion, snow avalanches, floods etc. are the examples of natural disasters while fire, epidemics, road, air, rail accidents and leakages of chemicals/nuclear installations etc. fall under the category of man-made disasters.

A disaster is an extreme disruption of the functioning of a society that causes widespread human, material, or environmental losses that exceed the ability of the affected society to cope with its own resources.

Disasters are sometimes classified according to whether they are *natural disasters*, or *man-made disasters*. For example, disasters caused by floods, droughts, tidal waves and earth tremors are generally considered *natural disasters*. Disasters caused by chemical or industrial accidents, environmental pollution, transport accidents and political unrest are classified as *human-made or human-induced disasters* since they are the direct result of human action.

A more modern and social understanding of disasters, however, views this distinction as artificial since most disasters result from the action or inaction of people and their social and economic structures. This happens by people living in ways that degrade their environment, developing and over populating urban centres, or creating and perpetuating social and economic systems. Communities and population settled in areas susceptible to the impact of a raging river or the violent tremors of the earth are placed in situations of high vulnerability because of their socio-economic conditions. This is compounded by every aspect of nature being subject to seasonal, annual and sudden fluctuations and also due to the unpredictability of the timing, frequency and magnitude of occurrence of the disasters.

Water and Climate-related Disasters

- Floods.
- Cyclones.
- Tornadoes and hurricanes.
- Hailstorms.
- Cloud bursts.
- Heat wave and cold wave.
- Snow avalanches.
- Droughts.
- Sea erosion.
- Thunder and lightening.
- Tsunami.

Geological Disasters

- Landslides and mudflows.
- Earthquakes.
- Dam failures/dam bursts.
- Minor fires.

Chemical, Industrial and Nuclear Disasters

- Chemical and industrial disasters.
- Nuclear disasters.

Accident-related Disasters

- Forest fires.
- Urban fires.
- Mine flooding.
- Oil spills.
- Major building collapse.
- Serial bomb blasts.

- Festival-related disasters.
- Electrical disasters and fires.
- Air, road and rail accidents.
- Boat capsizing.
- Village fire.

Biological Disasters

- Biological disasters and epidemics.
- Pest attacks.
- Cattle epidemics.
- Food poisoning.

Appendix 2

Kerala Floods 2018: Damage and Need Assessments

Spinal cord of the Kerala's economy has been conked out due the incessant and torrential rain which lasted for around 20 days in the month of August 2018. The state may take around 15 years to bring it into the mainstream of the economy. Kerala, one of the Indian provinces was submerged into water and witnessed the worst flood in the century. It was the worst flood after the Great Flood of 1924 in the state.

From recent news reports, it is reported that 483 people have died and 15 are missing (as on August 30, 2018). All the 14 districts of Kerala had been on the red alert. Millions of people have been evacuated from the affected area and are taking shelter in relief camps. According to newspaper reports of *The Hindu,* the Indian Government has declared it as a Level 3 calamity. According to the Kerala Government, one-sixth of the total population was directly affected by the floods. Most of the regions affected by this monsoon were classified as ecologically-sensitive zones (ESZs) by the Western Ghats Ecology Expert Panel, the Gadgil Committee. Most of the recommendations and directions by the committee were either neglected or rejected. Chairman of the committee Madhav Gadgil accused the state government and its irresponsible environmental policy for the recent landslides and floods.

Some of Prominent Factors Which Brought the Floods

The incident of flooding in the state is becoming more frequent due to heavy rainfall. Unchecked intervention of the resources is one of the main reasons of flooding in the state. According to BBC News, the rainfall was unusual; it was 257 percent more than the normal rainfall which Kerala receives during the monsoon. On mid-evening of August 8, 2018, the state received 12 inches of rainfall in the first 24 hours. It was for the first time in the history that 35 out of 54 dams were opened without a warning. The water level of the dams crossed the warning level due to the unexpected rainfall. A few of the dams were opened to release the excess water; and it was due to the communication gap among the dam authorities.

The State authority claims that opening of the Mullaperiyar Dam by the Tamil Nadu government was the only reason for floods in the state. Subsequently, Tamil Nadu Government rejected the allegations by stating that Kerala has suffered due to the sudden discharge of water from 80 reservoirs in the state. Roxy, Ghosh, et al., 2017, state in their research that rising temperatures have led to huge fluctuations in the monsoon winds carrying the moisture from Arabian Sea, resulting in heavy to extreme rains over the Western Ghats, lasting for two to three days.

Damage Assessment

There are two ways to assess the damages i.e. rapid damage assessment (RDA) and detail damage assessment (DDA). As per the RDA estimation, there was a ₹ 20,000 crore loss during the floods. Many people and livestock have died and property has been destroyed and damaged. Natural resources have been affected due to the floods. Millions of people have lost their livelihoods.

Common properties and infrastructure facilities have been heavily damaged. Moreover, the economy of the state has shrunk, tourist footfalls have also declined and many industries have been shut off. The consequences of the flood had impacted the state widely. They are mentioned below. Many people found their electrical appliances and furniture ruined and a thick layers of sludge and slime. Incidents of snake bites were reported from Ernakulam and Thrissur districts.

Death and Missing Cases of Humans

On August 30, 2018, it was reported that 483 people have died in the event and 15 people are still missing. The rescue team is working to save even the last person stranded.

Death of Livestock

The Animal Husbandry Department has released the data that the state has incurred loss of ₹ 110 crore due to the deaths of animals and birds. Approximately, 10,000 livestock have died due to the flood. More than 4 lakh of broiler chicken has also died in this event. It is estimated that nearly 50 percent of the cattle in flood-affected areas in Thrissur, Chengannur and Alleppey would have died and the Department is recording a similar loss in case of poultry and goats. Each district has a cattle population between 75,000-1,00,000, as per the AHD estimates. Deaths of cattle and poultry have also been reported from Idukki and Wayanad, where many landslides had occurred. "In many flood-affected areas, it was impossible for anyone to take the animals with them. Those animals which were let of might not have reached high grounds, as the extent of floods was huge", said an official with the AHD.

Of the 14 lakh cattle in the state, only around 2 lakh have been insured. The Department sets a target of 60,000 cattle every year and only milk yielding cows are included under the insurance coverage. The Department is now relying on unique identification process and geo-mapping for cows which was successfully carried out earlier to avoid possible disputes among farmers over ownership of cows. The scientific burial of animal carcasses is also being done after removing the eartag for future claims of compensation. A team comprising of the secretary of local body, livestock inspector, a veterinary doctor and health inspector of the local body have started monitoring the burial of carcasses in different areas.

Loss of Livelihoods

Millions of people have been temporarily displaced and residing in relief camps. All these people had to leave their livelihood activities back due to floods. The source of income for them have been halted for the time being.

Loss of Agriculture

As is a proven fact that floods damage agricultural crops, the flood in Kerala too has destroyed agriculture and has affected the farmers' economy. As per the report of State Agriculture Department, around 56,844.44 hectares of agricultural land has been destroyed causing a loss of ₹ 1,355.68 crore to 3.14 lakh farmers. The state is majorly a paddy production state which will lead to annual reduction in production of paddy in the country. Approximately, 26,106 hectares of paddy fields and 6,348 hectares of banana plants have been damaged. This is the estimated loss and the actual loss will be assessed further. There has also been a loss of soil fertility due to heavy flooding and continuous erosion. The high end areas have lost the fertility while the low lying areas have received the fertile soil from the upland due to the deposition of sediments. The fishing occupation of the fishermen in the state has been put on halt as all the fishermen were involved in the rescue operations.

The heaviest floods in Kerala since 1924 have taken a huge toll on the state's livestock and agriculture. Over 280,442 farmers in the state have been hit and their losses are estimated at ₹ 1,000 crore. Crop losses in over 45,988.50 hectares across the state, have been pegged at ₹ 10.7 billion. In 14 districts, over 114,000 acres of farms and orchards have been completely destroyed. Rice plantations have suffered the most. Approximately 45,000 acres of rice harvest has been gutted, over 23,000 acres of sago harvest, 1,500 acres of paddy plantations, 17,700 acres of coconut trees, 1,300 coconut plantations and 10,000 acres of vegetable harvest have been washed away. Around 75,000 pepper vines, 27,000 rubber trees, 15,000 nutmeg plants, 12,000 cocoa trees, 9,000 areca nut trees, 3,000 coconut trees and 1,455 coffee plants have been damaged due to the flood. Along with extensive losses of rubber plantations, spices like cardamom and pepper too have been washed away.

The initial estimated loss of agriculture is being pegged at ₹ 1,000 crore. Kerala is the country's largest producer of cardamom and rubber. Major damage to crops has been reported from Patthanamthitta, Alapuzha, Malappuram and Idukki districts. Idukki faces a loss of ₹ 1.3 billion and almost 28,000 farmers have been affected, quoted a senior state Agriculture Department official. The state gets 20 percent of its revenue from plantations. Over 18,316.38 ha of paddy, 7,161 ha of banana (bearing and non-bearing), 3,371 ha of cardamom, and 3,302 ha of vegetables have been damaged in Kerala, according to a primary report on crop loss by the state Farm Department.

Agricultural economist P. Indiradevi of Kerala Agriculture University said the loss estimated by the current system is incapable of fathoming the depth of the actual loss. For instance, the Department is assessing only crop loss, but there are various sectors that are directly or indirectly linked to the agriculture sector and the present system has no mechanism to assess the ripple effects of the crop loss in other related sectors.

Infrastructural Loss

The flood has damaged approximately 10,000 km. of road network in the state. Airline services and railways were also suspended temporarily. On August 29, 2018, Cochin International Airport resumed its service after the water was drained out from the runway. Thousands of homes have been destroyed. Many of schools and universities were suspended until the state returned to normalcy. Around 80 percent of the areas were blacked out so that there is no other mishap due to electric currents.

The water treatment plant of the state was forced to cease pumping water which resulted in restricted access to clean drinking water.

Damage to the Economy

The economy of the state has been standing still as the state is focusing on rescue and relief operations. The improvement of the economy will take some time to return to normalcy.

Rescue and Relief (RaR) Operations

After the floods, various departments came forward for rescue. There was a huge rescue operation by Indian Air Force and National Disaster Response Force (NDRF). Around 40 helicopters, 31 aircrafts, 500 boats and necessary equipment were put in service along with 182 different teams that were engaged in the rescue operations. This is considered to be one of the largest rescue operations in the disaster history of India. There were 18 medical teams of defence forces, 58 teams of National Disaster Response Force (NDRF), 7 companies of Central Armed Police Force (CAPF), 52 teams of central forces of Indian Navy and Indian Army and state governed forces like Kerala Police and Fire and Rescue Services that were involved in the service. Local fishermen also helped the victims to move to safer places or to relief camps. Human Society International helped rescuing the animals. More than 1.2 million people had taken shelter in more than 5,000 relief camps.

Many people have come forward to help the victims and the state in this crisis. Many government agencies, corporates, non-governmental organisations, individuals and foreign bodies are donating to help Kerala. The state had to cancel the celebration of Onam festival, instead using the funds for relief. The Kerala government on August 21, 2018 called for an all-party meeting to discuss the rehabilitation of people languishing in the relief camps. With rains easing and water receding in some areas of flood-ravaged Kerala, people started returning to their homes as Union Minister K.J. Alphons appealed for cooked food and skilled workers to rehabilitate more than 10 lakh displaced people. Though there was some respite from the rain, the water level did not come down in several areas of the worst-hit Ernakulam, Thrissur, Pathanamthitta, Alappuzha and Kollam districts, the officials said, a day after the Union Home Ministry declared the floods as *calamity of severe nature*.

Railways and road traffic services were restored in most places. The Minister, who belongs to Kerala, also issued an appeal to skilled workers like plumbers, electricians and carpenters to come to the state to make homes liveable and help in the task of restoring normalcy. Flagging the immediate needs of the people, he tweeted, "Doctors and nurses can go down to the villages as there is apprehension about the possibility of outbreak of deceases". As the state began the task of recovering from the disaster, more relief material arrived at the Cochin port. The state government, in an all-party meeting, elicited views on the package for rehabilitation and decided to honour the fishermen community that played a vital role in rescuing people.

The State Level Bankers Committee also decided to declare a moratorium on agriculture loan for one year. The income tax department has also advanced the time for filing returns for the Kerala tax payers.

The Department of Consumer Affairs had dispatched 100 MT of pulses. The Ministry of Health had airlifted 52 MT of emergency medicines. Another 20 MT of medicines also reached Kerala. 20MT of bleaching powder and one crore chlorine tablets were arranged for dispatch. At least 12 medical teams had been put on standby and no outbreak of disease were been reported so far.

The Ministry of Power had also extended help by moving power equipment like electricity meters, coils, transformers etc. to restore the distribution network. Power generation had reached 2600 MW and no difficulty was envisaged on this count.

The Department of Telecom made 77,000 towers operational out of a total 85,000. Out of 1,407 telephone exchanges, except 13, have been made functional. DoT had put a helpline with number 1948 in operation to help trace missing persons. The Ministry of Petroleum and Natural Gas made available 12,000 KL of kerosene. Adequate arrangements were made for distribution of LPG cylinders. Adequate aviation fuel was also made available.

With regard to cattle fodder, 450 MT was dispatched. Veterinary medicines had also been dispatched in two lots. Adequate quantities of milk and other products were also made available by the Department of Animal Husbandry and Dairying.

The NCMC meeting was attended by Secretaries of Central Ministries including Health, Telecom, Consumer Affairs, Food and Public Distribution, Power, Petroleum and Natural Gas, and senior officers of Defence Forces, Ministry of Home Affairs and the National Disaster Management Authority (NDMA).

Contributions from State Governments, Central Government, NGOs and Foreign Agencies

1. The Government of Kerala started a donation website named Kerala Chief Minister's Distress Relief Fund (CMDRF) to collect funds from all over the world. As on August 30, 2018, more than ₹ 700 crore had been received through this website.
2. Prime Minister of India announced ₹ 500 crore for the relief purpose in Kerala. More funds will be given as per the requirement in the near future for resettlement and rehabilitation.
3. The Home Minister of India also announced ₹ 100 crore for relief and rescue operations.
4. NDRF also committed to release funds after the detailed damage assessment is done.
5. Many Members of the Parliament, government employees, Members of the Legislative Assembly and civil servants across the country donated a large amount through the

CMDRF.

6. State governments donated a sum of ₹ 211 crore to help the victims and also arranged for sending the relief material like packaged water, blankets, packaged food, rice, healthcare products etc.
7. European Union through Indian Red Cross Society donated ₹ 1.53 crore for the relief operations.
8. Other countries like Maldives and Thailand also supported with aids and supplies.
9. Israel-Aid, Israel sent relief workers to distribute relief supplies.
10. Many local NGOs also worked to help the victims in all the possible ways.
11. Bihar Chief Minister Nitish Kumar announced an aid of ₹ 10 crore.
12. Chief Minister of Orissa Shri Naveen Patnaik announced an aid of additional ₹ 5 crore to the earlier sanctioned ₹ 5 crore along with 245 fire personnel with boats.
13. Similarly, contributions from Tamil Nadu (₹ 10 crore), Andhra Pradesh (₹ 10 crore), Puducherry (₹ 1 crore), Jharkhand (₹ 5 crore), Maharashtra (₹ 20 crore), Gujarat (₹ 10 crore), Punjab (₹ 10 crore), Telangana (₹ 25 crore), Bihar (₹ 10 crore), Himachal Pradesh (₹ 5 crore), Uttarakhand (₹ 5 crore), Chattisgarh (₹ 3 crore) Madhya Pradesh (₹ 10 crore), Karnataka (₹ 10 crore), West Bengal (₹ 10 crore), Manipur (₹ 2 crore) and Tripura (₹ 1 crore) were announced as a part of donations.
14. While Congress demanded the disaster in Kerala to be termed as 'national calamity', former Jammu and Kashmir Chief Minister Omar Abdullah said he would donate a month's salary to the flood-ravaged state and appealed to his colleagues in the state also to do the same. Tamil Nadu IAS officers pledged a day's salaries towards the flood relief in Kerala. The Islamic Centre of India urged Muslims to donate 10 percent of Eid budget to Kerala flood victims.

15. AMMA, the organisation which was under scrutiny recently in Kerala donated ₹ 10 lakh whereas Malayalam actors Mammooty and Dulquer Salmaan contributed ₹ 15 lakh and ₹ 10 lakh respectively. Malayalam film industry superstar Mohanlal gave ₹ 25 lakh to the CMDRF, whereas, actor Jayasurya distributed essentials to the inmates in relief camps in Paravur.
16. Pakistan Prime Minister Imran Khan, while expressing solidarity in his tweet, offered to provide humanitarian assistance. The governments of Maldives and Thailand have also offered aid and support.
17. Chief Ministers of almost all the states and Union Territories pledged monetary aid from their respective state funds in addition to dispatch of various relief materials such as potable water, blankets, packed food, rice, water-purifying machines, daily-use and healthcare products.

Contributions from Corporates and Individuals

1. Many e-commerce websites like Facebook, Amazon, Flipkart and Paytm provided a link for donation on their portals.
2. Major oil and gas companies like IOCL, HPCL, Oil India Limited, ONGC collectively donated ₹ 25 crore through CMDRF and also provided relief materials.
3. National Dairy Development Board (NDDB) provided the state with huge quantities of tetra pack milk.
4. Mrs. Nita Ambani, Chairperson of Reliance Foundation also announced a sum of ₹ 21 crore through CMDRF and also relief material worth ₹ 50 crore.
5. Adani Foundation CSR also committed to provide ₹ 1 crore for immediate relief and ₹ 1 crore for resettlement and rehabilitation.
6. 21st Century Fox and their employees collected ₹ 5 crore and donated towards the relief operations.
7. Bollywood industry also contributed to the CMDRF.

8. Mr. K.P. Hussain, Chairman of Fathima Healthcare Group, also donated ₹ 1 crore for relief and another ₹ 4 crore for health relief aid.
9. Dr. Shamsheer Vayalil announced to donate an amount of ₹ 50 crore for the relief purpose.
10. A UAE businessman donated a sum of ₹ 18.85 crore.
11. Canara Bank donated ₹ 5.01 crore through CMDRF along with 10 tonnes of rice.
12. Indian Cricket team also donated one match fee towards the calamity.
13. The Cochin Port Trust had earmarked two warehouses for free storage of relief materials, arriving from across the country, for people affected by the unprecedented floods.
14. Various businessmen and individuals also donated monetary funds to help the relief work in Kerala including NRI businessman M.A. Yusuf Ali who donated ₹ 5 crore. The recent 21-year-old student from Kerala who was trolled for selling fishes, Hanan also donated ₹ 1.5 lakh towards the Chief Minister Relief fund that she had received from people to fund her studies and take care of her family. Tamil Nadu political party DMK donated ₹ 1 crore as well.
15. BJP MP Varun Gandhi donated ₹ 2 lakh to the CMRDF and urged other politicians to also contribute to the fund.
16. Attorney General K.K. Venugopal donated ₹ 1 crore while a Kerala pastor from Pathanamthitta district offered 25 cents of his land to bury those dead in the floods.
17. State Bank of India (SBI) donated ₹ 2 crore for the Kerala flood victims apart from ground-level measures to help them. They have also waived off fees and charges on services like loans for flood relief, duplicate passbooks, ATM cards, cheque books and EMI delays, besides making attempts to restore the working of branches and ATMs in the flood-affected regions of the state.
18. Various donation camps also sprung up across the country in various cities like New Delhi, Mumbai, Chennai and

Bengaluru. People were toiling through the day to collect and segregate funds for the flood-ravaged state which included monetary relief as well as funds in the form of supplies and stock. Indian Railways announced free transportation for relief material across to Kerala as well. The Department of Food and Public Distribution has provided 50,000 million tonnes (MT) of food grains for the victims so far.

19. XISS, Ranchi which is a service based educational institute donated one day salary of all employees of the organization for CMDRF.

Immediate Response

The State and the Central government put their best efforts to control the situation. The rescue teams responded quickly to help the victims in all possible ways. The fisherman helped a lot in the operation for which the state government announced rewards for their bravery and brotherhood. International agencies also responded by sending teams for the rescue and relief operations.

Preparedness Plan Taken by the Government

Early Warning and Emergency Communication System: A VHF based early warning system is installed in 295 vulnerable villages in the state with the support of Asian Development Bank and United Nations Development Programme in order to strengthen the emergency response communication capabilities and early warning facilities of the State and District administrations and to enhance coordination between State, District administrations and villages. This wireless radio system offers a free flow of communication between district headquarters and village offices/taluk offices. The wireless radio installed in the villages is connected to a PA system through an amplifier, facilitating communication from district headquarters to the general public. The PA system installed at the top of the village office building can disseminate warnings or messages from district administration office to a distance of 1.5 km. (radius), enabling the district collector to directly communicate with the people living in these villages. Other than disseminating warnings, the system also provides communication facilities between district headquarters and village offices/taluk offices facilitating coordinated response and relief in the event of a disaster.

Community Level Preparedness: The Government of Kerala has implemented a Community Based Disaster Risk Reduction (CBDRR) Programme in 197 vulnerable villages covering nine coastal villages with the support of United Nations Development Programme that aimed at building local capacities at various levels and promoted preparedness. Formation of community-based hazard prevention, mitigation and preparedness plans and constitution of disaster management committees and task forces were the main components of the project.

School Suraksha Programme: Department of Revenue and Disaster Management has initiated a 'School Suraksha Programme' in association with Department of Education in order to generate awareness on disasters at the school level, to promote a culture of disaster preparedness in the school community in the state and to promote disaster management clubs/teams in schools. The programme launched in 2010 will cover all the government schools in the State. ₹ 1.75 crore was allocated by the Department of Revenue and Disaster Management for the implementation of the programme.

Rapid Need Assessment

Food Security and Livelihoods: The major assessment findings across all districts are:

- There is an urgent need for food aid as well as for logistical means to carry out the distribution of food and other relief items in the flooded areas. In order to prevent a deterioration of the nutritional status of the most vulnerable populations, these families require continued relief assistance.

- Providing food items is not only for the people sheltered in relief camps but also for those who will be returning to their homes once the water levels recedes. Most of the vulnerable communities including tribal communities have lost their main source of income. Due to the loss of harvest in the rural areas, in a period of two to three months from now, this lack of food production might generate a deficit in the food security conditions and have a potential negative impact on coping mechanisms at the household level.
- Furthermore, agricultural activities have been severely affected by the floods. It is, therefore, important to help restore livelihoods through the promotion of short-term emergency agricultural activities (distribution of seeds, agricultural inputs and tools). Most of the farmers plant paddy seeds during the monsoon period and the flood water has damaged most of the agricultural fields.
- There is severe shortage of cattle fodder. Most of the grazing lands were affected with submerged flood water. Most of the ponds were also flooded.

Water Sanitation and Hygiene (WASH): The major assessment findings across all districts are:

- The people of the area used to drink from wells and interior village people from the stream water.
- Some wells have been damaged or are contaminated. Presently, people are managing from the stream water.
- In the camps, people are getting safe drinking water.
- Families are in camps where sanitation and toilet facilities are available separately for males and females.
- Pregnant, lactating mothers, common women and adolescent girls are managing well to care of their personal hygiene.

Shelter Situation Analysis:

- In most of cases, affected families were able to retrieve the documents and valuables.

- The *kuccha* old houses (made of mud and cement) were the most vulnerable and are badly damaged. Poor and middle class both were affected but the middle class families opined that they will manage; hence, extending support to poor families should be the priority, especially the shelter support.
- Government is proactive which can be seen through the compensation they have announced plus the deadline set for releasing the amount in affected families' bank accounts.
- Local volunteers, NGOs, political parties, Rotary members, local faith-based organizations, all came forward to support and they have also formed Palakkad relief team which is working non-stop. Immediate needs of the people have been met.
- Panchayat level officials were active for debris clearance. The Collector's office had put advertisements in the newspaper for a NGO coordination cell at the Collector's office and they received more than 300 volunteers who were deputed in the affected villages.
- All the camps are well managed by the community members voluntarily especially those who are not affected by floods. The local people started supplies of dry rice, dal, vegetables, clothes and even generously donated cash etc.
- Allopathic, homeopathic and ayurvedic medical centres have been set up in each camp. Mostly, people suffering from fever, cough and cold, and hyper tension are coming for treatment as per the information shared by the medical teams.
- Ambulance has been provided in each camp in case of any emergency hospitalisation.

Protection: Assessment Findings:

- Men, women, adolescent girls, children are living together inside the camps. There is separate staying arrangement for males and females.

- Women and adolescent girls are managing well their personal hygiene.

Education:

- Most of the children have lost their education material. Hence, there is an urgent need for providing study material including uniforms.
- Government may take the responsibility to issue certificates and other credentials to the students who have lost their documents in the floods.

Recommendations for Rehabilitation

- Food and livelihood.
- Agricultural support.
- Livestock shelter.
- Construction of temporary houses.
- Support for repairing the damaged houses.
- Hygiene kits (sanitary pads, soaps, menstrual hygiene materials, nail cutters etc.)
- Public health promotion.
- Unconditional cash Transfer (UCT).
- School dress and books for the school going children.

(SPHERE India Report, 2018, has been used with thanks to make sufficient portion of this chapter)

Bibliography

Bibliography

Asthana, Vandana (1992), "The Politics of Environment: A Profile", Ashish Publishing House, Delhi.

Bahuguna, Sunderlal, Vandana Shiva and M.N. Buch (1992), "Environment Crisis and Sustainable Development", Natraj Publishers, Dehradun.

Bose, Ranjan K., and V. Srinivasachary (1997), "Policies to Reduce Energy Use and Environmental Emissions in the Transport Sector: A Case of Delhi City", *Energy Policy,* Vol. 25, 1137-1150.

Centre for Science and Environment (CSE) (1990), Proceedings of the Seminar on Traditional Water Harvesting System in India, October 9-11, New Delhi.

Dasgupta, Biplab (1978), "The Environment Debate: Some Issues and Trends", *Economic and Political Weekly*, Annual.

Dogra, Bharat (1991), "Environment Protection: Myths and Reality", *The Financial Express*, New Delhi, July 27.

Gautham, Naviakha (2000), "Urban Pollution: Driving Workers to Desperation", *Economic and Political Weekly,* 35(51), December 16-22.

Government of India, Ministry of Environment, Forest and Climate Change, *Annual Report, 2015-16.*

—Ministry of Environment and Forest, *National Environmental Policy*, 2006.

—Ministry of Environment and Forest, *National Report on Implementation of the United Nations Convention to Combat Desertification,* 2006.

—Ministry of Environment, Forest and Climate Change, Forest Survey of India, *State of Forest Report* (various years).

—Ministry of Environment, Forest and Climate Change, Central Pollution Control Board, *Annual Report* (various years).

Guha, Ramchandra (1988), "Ideological Trends in Indian Environmentalism", *Economic and Political Weekly*, Vol. 23, No. 49, December 3.

Guha, Ramchandra (1989), "The Unquiet Woods: Ecological Change and Peasant Resistance in the Himalayas", Oxford University Press, New Delhi.

Hanumantha, Rao C.H. (2000), "Watershed Development in India: Recent Experience and Emerging Issues", *Economic and Political Weekly*, Vol. 35, No. 45, 3943-3948.

Jadhav, H.V. (1995), "Environmental Pollution", Himalaya Publishing House, Bombay.

Jerry, A. Nathanson (2002), "Basic Environmental Technology", Fourth Edition, Prentice Hall of India Pvt. Ltd., New Delhi.

Kathuria, Vanish and G.S. Haripriya (2000), "Industrial Pollution Control: Choosing the Right Option", *Economic and Political Weekly*, October 28.

Krishna, Kannan (1997), "Fundamental of Environmental Pollution", S. Chand and Co., New Delhi.

Lal, J.B. (1989), "India's Forests: Myth and Reality", Natraj Publishers, New Delhi.

Luthra, A. (1999), "Vehicular Emissions in India: Retrospect and Prospects", *Environment and People,* Vol. 16 (4).

Mall, R.K., R. Kumar and R. Bhatla (2010), "Climate Change and Disasters in India", *Journal of South Asia Disaster Studies*, SAARC Disaster Management Centre, New Delhi.

Nagdeve, D. (2004), "Air Pollution in Mega Cities of India", International Institute for Population Sciences.

Omvedt, Gail (1984), "Ecology and Social Movements", *Economic and Political Weekly*, November 3.

Pitman, G.T. Keith (2002), "India: World Bank Assistance for Water Resources Management: A Country Assistance Evaluation", World Bank.

Rai, Usha (1992), "Saving the Environment", *The Indian Express*, New Delhi, March 30.

Ramesh, R. and M.G. Yadava (2005), "Climate and Water Resources of India", *Current Science,* Vol. 89(5).

Ravindranath, N.H. and J. Sathaye (2002), "Climate Change and Developing Countries", Kluwer Academic Publishers, Netherlands.

Reducing Disaster Risk: A Challenge for Development by UNDP (BCPR), 2004.

Report of the High Powered Committee on Disaster Management in India, National Centre for Disaster Management, Indian Institute of Public Administration, New Delhi.

Report of Working Group on Disaster Management for the Eleventh Five Year Plan (2007-2012), Planning Commission, December 2006, New Delhi.

Rupa, Kumar K. et al. (2005), "High-resolution Climate Change Scenarios for India for the 21st Century", *Current Science*, Vol. 90, 334-345.

Sapru, R.K. (ed.) (1987), "Environmental Management in India", Ashish Publishing House, New Delhi.

Sathaye, J., P.R. Shukla and N.H. Ravindranath (2006), "Climate Change, Sustainable Development and India: Global and National Concerns", *Current Science,* Vol. 90(3).

Sidhu, B.S. and V. Beri (1989), "Effects of Crop Residue Management on the Yields of Different Crops and Soil Properties", *Biological Wastes*, 27, 15-27.

Singh, Ram Bali (ed.) (1988), "Study in Environment and Development", Commonwealth Publications, New Delhi.

Singh, Shekhar (1985), "Some Aspects of Ecological Crisis in India", *Social Scientist.*

Websites

Ministry of Environment, Forest and Climate Change

Ministry of Water Resources, River Development and Ganga Rejuvenation

Ministry of Mines

Ministry of Coal
Ministry of Petroleum and Natural Gas
Ministry of Health and Family Welfare
NITI Aayog

Index

Index

A

Adaptation Fund Board (AFB), 56
Agricultural Drought, 72

B

Bombay Public Trust Act, 1950, 98
Business-to-business Good Practices, 119
Calamity Relief Fund (CRF), 25

C

Causes of Recurrence of Droughts in India, 71
Central Board of Secondary Education (CBSE), 103
Classification of Droughts, 72
Climate Change and Human Settlements, 40
Climate Change and Water Supply, 41
Coal Cess, 11
Community-based Disaster Preparedness (CBDP), 22
Companies Act, 2013, 138
Corporate Social Responsibility Voluntary Guidelines, 138
Corporate Social Responsibility, 112
Crop Weather Watch Group (CWWG), 78

D

Disaster Financial Assistance Arrangements (DFAA), 145
Disaster Management Act, 2002, 146
Disaster Management Act, 2005, 90
Disaster Management and Development Planning, 89
Disaster Management and Finance Commissions, 95
Disaster Management Education in India, 103
Distribution of Drought-prone Areas, 75
Duration of the Drought, 74

E

Earth Summit, 12
Eco-Industrial Hubs, 12
Eco-management and Audit Scheme (EMAS), 123

Eighth Five Year Plan (1992-1997), 85
Eleventh Five Year Plan (2007-12), 84
Emergency Management Australia (EMA), 145
Environmental Taxes, 11
Equator Principles, 136
Expert Group on Low Carbon Strategies for Inclusive Growth, 60

F

Floods in Bangladesh, 25

G

G.N. Ritchie, 102
Gas Authority of India Limited (GAIL), 36
Global Dimension of CSR, 127
Global Reporting Initiative (GRI), 136
Green India Mission (GIM), 52
Green Revolution, 67
Gujarat Institute of Disaster Management (GIDM), 96
Gujarat Pollution Control Board, 36
Gulf of Khambhat, 36

H

Hazard Mitigation Grant Programme (HMGP), 144
Human-Made Disasters, 149
Hurricane Katrina, 107
Hydrological Drought, 72
Hyogo Framework for Action, 8

I

India Disaster Knowledge Network, 105
Indian Coast Guard, 36
Indian Institute of Public Administration (IIPA), 106
Indian Network for Climate Change Assessment (INCCA), 58
Industrial Effluents, 65
Infant Death Syndrome Prevention (IDSP), 107
Intensive Agricultural Practices, 64
Inter-Governmental Panel on Climate Change (IPCC), 45
International Union for Conservation of Nature (IUCN), 15

J

Jawaharlal Nehru National Solar Mission (JNNSM), 48

K

Kyoto Protocol, 136

M
Main Features of CSR, 111
Meteorological Drought, 72
Millennium Development Goals, 13
Ministry of Environment, Forest and Climate Change (MoEFCC), 47
Ministry of Human Resource Development (HRD), 103
Ministry of Petroleum, 34
Montreal Protocol, 16

N
Narmada River, 36
National Action Plan on Climate Change (NAPCC), 47
National Calamity Contingency Fund (NCCF), 95
National Carbonaceous Aerosols Programme (NCAP), 59
National Centre for Disaster Management (NCDM), 106
National Civil Defence Secretariat, 146
National Clean Energy Fund (NCEF), 55
National Disaster Management Act 2005, 18
National Disaster Management Authority (NDMA), 91
National Disaster Response Force (NDRF), 92
National Institute for Climate Change Studies and Actions (NICCSA), 60
National Institute of Disaster Management (NIDM), 93
National Mission for Enhanced Energy Efficiency (NMEEE), 49
National Mission for Sustainable Agriculture (NMSA), 53
National Mission for Sustaining Himalayan Ecosystem (NMSHE), 51
National Mission on Strategic Knowledge for Climate Change (NMSKCC), 50
National Mission on Sustainable Habitat (NMSH), 53
National Water Mission (NWM), 52
Natural Disaster Relief and Recovery

Arrangement (NDRRA), 145
Natural Disasters, 149
Nigerian Dilemma, 37
Non-monetary Incentives, 12

O

Occurrence of Droughts, 42
Oil and Natural Gas Corporation (ONGC), 36
Origins of CSR, 111
Orissa State Disaster Management Authority (OSDMA), 104
Orissa Super Cyclone, 19

P

Pipeline Transportation, 33
Polluter Pays Principle, 11
Prime Minister's Council on Climate Change (PMCCC), 57
Principle of Common but Differentiated Responsibility, 15
Principle of Triple Bottom Line (TBL), 116
Programmes for Treatment of Degraded Lands, 81

R

R. Edward Freeman, 112
Reclamation and Development of Alkali and Acid Soils (RADAS), 84
Reserve Bank of India (RBI), 135
River Hydrology, 43

S

Sea Level Rise, 44
Sendai Framework for Disaster Risk Reduction, 9
Social Responsibilities of Businessmen, 111
Socially Responsible Investment (SRI), 123
Societies, Registration Act, 1860, 98
Soil and Land Use Survey of India (SLUSI), 82
Soil Conservation Training Centre, 83
State Action Plan on Climate Change, 55
State and District Disaster Management Authorities (SDMAs), 92
Subsidies, 11
Sustainable Development Goals (SDGs), 2
Sustainable Development, 1

T

Tenth Five Year Plan

(2002-07), 23
Twelfth Five Year Plan (2012-17), 93

U
U.N. Framework Convention on Climate Change (UNFCCC), 15
United Nations Conference on Environment and Development (UNCED), 13
United Nations Environment Programme (UNEP), 15
United Nations Environment Programme-Finance Initiative (UNEP-FI), 131
Unsanitary Disposal of Solid Wastes, 64

V
Vector Borne Diseases, 30

W
Washington Post, 107
World Business Council for Sustainable Development (WBCSD), 142
World Conference on Disaster Risk Reduction, 4
World Health Organization (WHO), 30